Quarterly Essay

Quarterly Essay is published four times a year by Black Inc., an imprint of Schwartz Books Pty Ltd. Publisher: Morry Schwartz.

ISBN 9781760644390 ISSN 1444-884x

Subscriptions – 1 year print & digital
(4 issues): $89.99 within Australia incl. GST.
Outside Australia $124.99. 2 years print & digital
(8 issues): $169.99 within Australia incl. GST.
1 year digital only: $59.99.

Payment may be made by Mastercard or Visa, or by cheque made out to Schwartz Books. Payment includes postage and handling.

To subscribe, fill out and post the subscription card or form inside this issue, or subscribe online:

quarterlyessay.com
subscribe@quarterlyessay.com
Phone: 61 3 9486 0288

Correspondence should be addressed to:

The Editor, Quarterly Essay
22–24 Northumberland Street
Collingwood VIC 3066 Australia
Phone: 61 3 9486 0288 / Fax: 61 3 9011 6106
Email: quarterlyessay@blackincbooks.com

Editor: Chris Feik. Management: Elisabeth Young. Publicity: Anna Lensky. Design: Guy Mirabella. Associate Editor: Kirstie Innes-Will. Production Coordinator: Marilyn de Castro. Typesetting: Typography Studio.

Printed in Australia by McPherson's Printing Group. The paper used to produce this book comes from wood grown in sustainable forests.

HIGHWAY TO HELL

Climate change and Australia's future

Joëlle Gergis

Late in the summer of 2023–24, I found myself having a drink with a senator on a rainy afternoon in Canberra. It's not every day a climate scientist like me gets to talk directly to a politician about my work. As someone who spends most days in the neat cocoon of my research world, I have always found the rough and tumble of politics intimidating. Whenever I watch the news, parliamentarians seem to be shouting at each other, polarising important conversations about the issues that affect us all. Our leaders are becoming more tribal, resulting in an aggressive, highly adversarial form of politics that is demoralising and alienating more and more people each day.

Few topics have divided Australians more than climate change. Because we have always had dramatic swings in our weather, some people think that the increasingly destructive climate we are now experiencing is simply part of a "natural cycle." Apparently we've been through it all before; it's just a bunch of greenies trying to scare our kids. This opinion doesn't sit well with the evidence amassed by the world's leading scientists, which shows "human influence on the climate system is now an established fact." The proof is now so indisputable that it's like saying the sky is blue or the Earth is round. Whether or not everyone understands the science behind global warming is irrelevant – our climate is changing because of what humans have been doing to the planet throughout the entire course of our history.

Despite Australia being one of the most climate change–exposed countries in the world, our political response to addressing the issue has a long and chequered history of delay and denial. We are in the embarrassing position of being the only nation to have successfully introduced a carbon price and then revoked it with a change of government. After its introduction in 2012 by the Labor government led by Julia Gillard, Tony Abbott repealed the scheme in 2014 and restated his support for the fossil fuel industry, saying coal is "good for humanity." Progress in Australia's climate policy often looks like a case of one step forward and two steps back.

For most people just trying to get on with their lives, thinking about climate change feels too complicated and overwhelming, especially in an era of war, pandemic and financial pressure. But the problem is that governments around the world are making fateful decisions *right now* that will shape our planet's future. Meanwhile, most Australians aren't aware how bad things are and how much worse they will get. But when we tune out, we squander the most powerful thing we have to influence our society: our vote. In 1959 Martin Luther King Jr summarised the harm of political apathy by saying: "If you fail to act now, history will have to record that the greatest tragedy of this period of social transition was not the strident clamour of the bad people, but the appalling silence of the good people."

Since serving as a lead author on the latest United Nations Intergovernmental Panel on Climate Change (IPCC) report, I've become increasingly vocal about the threats Australia now faces. Over the past few years I've done hundreds of media interviews, public talks and festival events, published two books and a long list of newspaper and magazine articles trying to warn the public. But every time, I've been careful not to weigh in on political matters, for fear my colleagues would think less of me. Although it's possible to be rational and professional while being honest and humane, science is a very conservative profession that values detached objectivity above all else. But as the planetary crisis escalates, the ethical toll of silence now feels unbearable. As I wrote in *The Monthly*: "When experts fail to engage authentically in public conversations about climate change, others will step in to fill the silence. Commentators unconstrained by the professional ethics and rigour of our discipline have generated rife misinformation that has led to the shameful complacency plaguing the political response to the climate change problem for decades."

So this is how I came to find myself sitting across the table from one of the most influential politicians in the country on a Wednesday afternoon, instead of across town teaching my yearly batch of 200 university students. The senator contacted me after reading my book *Humanity's Moment* and wanted to know what he could do to help at this fraught time in our political history. As a scientist, I wanted to hear, first-hand, what is going on in Parliament House that is stopping our leaders from addressing climate change in a rational way.

He winced as I updated him on the latest science and shared my fears for the future of our nation: an endless list of broken records, signs of abrupt climate change; parts of our country becoming uninhabitable within our lifetime. He confided equally terrifying accounts of how the fossil fuel industry has a stranglehold on our federal government: bearing witness to the passage of legislation to appease lobbyists; to parliamentarians indignantly arguing a case against the government's duty of care to provide future generations with a safe climate; and to the disgraceful behaviour of self-serving political animals undermining the national interest.

As a newcomer to politics, he said he didn't understand "how they sleep at night" knowing that they are making decisions that will lock in warming for centuries to come. It was a meeting of kindred spirits. There we were, an unlikely pair – a politician genuinely willing to listen and a climate scientist trying to understand what our leaders and communities need to do to turn the situation around. By the end of our conversation, it was clear to us both that climate change is not just a scientific problem – it is the great moral challenge of our time. The essay you are holding in your hands is a guide to navigating the ethical dilemma we all must face about the future of our nation.

As we gathered up our belongings and prepared to dash out into the rain, we acknowledged that the next federal election could be the most decisive in our history: the strength of our government's climate policy will make or break Australia's future prosperity and influence the liveability of the entire planet for centuries to come. As Australia is the third-largest exporter of fossil fuels in the world, what we do over the next handful of years really, really matters. The time for standing on the sidelines has passed. All of us must use our vote to deliver more people to our parliament who are willing to fight for environmental protection, intergenerational justice, social equity and a sustainable economy. What happens next is up to all of us.

*

On 21 May 2022, after a decade of deep inertia in Australian climate politics, the nation found itself at a crossroads. Under Scott Morrison's leadership, the Coalition proposed to cut greenhouse gas emissions by 26 to 28 per cent from 2005 levels by 2030 and promised a "gas-led recovery" out of the economic downturn caused by COVID-19. The Coalition did not hide its plans to exploit fossil fuels beyond 2050 – a time when the world needs to reach net zero emissions to stabilise the Earth's climate. Australia's fossil fuel industry could continue with "business as usual" by relying heavily on carbon-offset schemes and the use of carbon capture and storage (CCS) technology. We will come back to this later, but for now you can effectively think of the Coalition's climate policy as a plan to keep burning fossil fuels into the foreseeable future while trying to soak up the industry's excess carbon by planting trees and promising not to clear land, or trying to bury the waste underground using non-existent technologies. The status quo would prevail.

In contrast, the Australian Labor Party (ALP) offered to listen to voters' deep concern about our increasingly destructive climate, proposing an alternative to the decade of delay and obstructionism engineered by the Coalition. Labor's relatively more ambitious suite of policies promised a 43 per cent reduction in emissions from 2005 levels by 2030 with the goal of achieving net zero by 2050; substantial investment in renewable energy technology, including increasing the share of renewable energy in the National Electricity

Market to 82 per cent by 2030; and adjustments to the Safeguard Mechanism – a 2016 Coalition policy to encourage reductions in emissions from Australia's largest polluters – to introduce a tighter cap on industrial emissions. After years of bitter political debate, Labor assured disillusioned voters that it would end the "climate wars." But when asked whether it supported the continued development of fossil fuel projects, it remained evasive. After years in the political wilderness, this was a make-or-break election Labor couldn't afford to lose.

Consequently, the ALP's *Powering Australia* plan did not mention coal or gas once. When questioned by journalists, Anthony Albanese said his party would support new coal projects if they "stack up environmentally, and then commercially." The plan made oblique reference to the fact that "our largest exporters know that global markets are moving to a low-carbon future." This signalled quiet support for the gas industry as a less carbon-intensive "transition fuel," while giving the public the impression that, under Labor leadership, Australia was poised to become a climate hero – we would swoop in and become the clean energy superpower needed to save the planet.

In the aftermath of the Black Summer bushfires of 2019–20 and the catastrophic flooding of eastern Australia in the early months of 2022, people all over the country were weary. We had a hell of a time getting through the health and economic battering of the COVID-19 pandemic, as a string of "once-in-a-lifetime" weather extremes unfolded. Many people were brought to their knees: fires burnt down their homes; floodwaters caked everything they owned in mud. Again and again and again. When things hit rock-bottom, people felt enraged by the government's shambolic response, which left them having to rely on volunteers from their local communities in their hour of need. The last few years had been a shameful display of successive governments' inability to comprehend scientific warnings and their failure to adequately prepare the nation for a changing climate. So by the time the 2022 federal election rolled around in May, although there were gaping holes in Labor's plan, *anything* was better than more of the same from the Coalition.

Like many on election night, I stayed awake until the early hours of the morning, watching the blow-by-blow coverage on television. As an IPCC author, I am one of the few people who understands the scientific reality of what the UN secretary-general, António Guterres, means when he says: "We are on a highway to climate hell with our foot still on the accelerator."

In a defining moment in Australian political history, the Morrison government was emphatically voted out. I wept watching an emotional Anthony Albanese deliver his victory speech, promising to do better. I believed him. Although the ALP had won, there had been a seismic shift in Australian politics away from the traditional two-party system. An extraordinary ten parliamentary seats in the House of Representatives and one in the Senate were won by independent candidates campaigning strongly on climate action. The phenomenal success of their grassroots campaigns took the major political parties by surprise, as previously safe conservative seats were swept away in an unprecedented groundswell of community support for independents all over the country. The Australian Greens also recorded their strongest ever result, claiming four seats in the lower house and twelve in the Senate, strengthening the crossbench support needed for stronger climate policy. It seemed as though the long-overdue "climate election" that disillusioned voters had been waiting for had finally arrived.

Although the 2022 federal election ushered in a new era of progressive politics in Australia, as Labor's first term in power has progressed, many people are now wondering if the political deadlock on our nation's climate policy has really been broken. Although some good ground has been made, particularly in legislating Australia's net zero by 2050 emissions target, scaling up support for renewable energy and placing tighter restrictions on industrial emissions, the federal government's actions still don't reflect the urgency of the planetary-scale crisis we are in. Australia's greenhouse gas emissions are rising and enormous fossil fuel projects continue to be approved to meet domestic and international demand.

While any reasonable person understands that Rome wasn't built in a day, the truth is that we are still not doing enough to address the root cause of our rapidly warming planet. The IPCC – the world's authority on climate science – has very clearly demonstrated that the burning of fossil fuels is the primary cause of climate change. Virtually all the observed global warming we have experienced since the Industrial Revolution has been driven by emissions from human activities. The latest figures from the Global Carbon Project show that 82 per cent of carbon dioxide emissions since 1960 have come from the burning of coal, oil and gas, with the remaining 18 per cent generated from land use changes like deforestation and the degradation of ecosystems. And the problem is getting worse: total anthropogenic emissions more than doubled over the last sixty years, with almost half of the carbon dioxide accumulated in the atmosphere emitted since 1990. In 2022, the burning of coal was the largest contributor to global carbon emissions (41 per cent), followed by oil (32 per cent) and gas (21 per cent). No matter which way you look at it, fossil fuels are cooking the planet.

The scientific reality is, regardless of political spin used to justify the continued exploitation of fossil fuel reserves, the laws of physics will keep warming the planet until we stop pumping carbon into the atmosphere and begin cleaning up the mess. The situation is too far gone for renewable energy alone to save us. Pinning our hopes on carbon capture technology to justify the continued burning of fossil fuels is a disastrous gamble the world can't afford to take. So, as this fateful moment approaches, we need to take an honest look at the government's climate policy and realistically assess the situation we are in. Are the climate wars really over, or has a new era of greenwashing just begun?

*

Events of recent years have graphically showcased how underprepared Australia is for the impacts we've already experienced with 1.2°C of global warming, let alone the conditions we will experience in the future. As I will explain, there is a 90 per cent chance that the continuation of current climate policies will result in 2.3°C to 4.5°C of global warming by the end of century, with a best estimate of 3.5°C. This represents a catastrophic overshooting of the Paris Agreement targets, highlighting just how far off track we really are. The lower limit of 1.5°C is expected to be breached in the early 2030s, with 2°C reached in the 2040s – within the lifetime of most people alive today. The consequences of such high levels of warming on the Australian way of life and on our national security, health and unique ecosystems are profound and immeasurable. To force our political leaders to do better, we need to wake up to the fact that we are the last generation that will experience the world as we know it today: the last to experience the tropical wonders of the Great Barrier Reef, summers uninterrupted by life-threatening storms, and the awe of wild places before they succumb irrevocably to the ravages of fire and intolerable heat.

Australia is in peril, and yet the consequences of a warming planet on our sunburnt country are still poorly understood by most people outside of the scientific community, let alone by our government. In 2023 a national study of just over 4000 Australians was published by Queensland's Griffith University showing a major disconnection between the scientific reality of climate change and the public's perception of the severity of the problem. Although three-quarters of Australians surveyed accept that climate change is real, only 15 per cent think it is an "extremely serious" problem right now. The poll showed that while close to a third of people believed climate change will be an issue in 2050, the urgency of addressing the problem was not appreciated. The most interesting result for me was that around a quarter of Australians do not understand or accept the science and the reality of the threats posed by climate change, identifying as deniers, sceptical or unconvinced. I would have expected this a decade ago, but surely these results don't reflect the reality of today – or do they?

To get a sense of how representative the Griffith University study might be, I tracked down a 2023 global survey of views on climate change. To my surprise, this polling also showed a disturbing lack of awareness of the scientific reality of climate change – over half of the Australians surveyed claimed that the impacts in our region have not been severe, with a third of people believing that the media exaggerates the influence of global warming. These perceptions may help explain why governments remain committed to only advancing modest climate policies that have broad electoral appeal. People are willing to install solar panels on their roof if it saves them money on their power bills, but they switch off when the harder conversation around the urgency of shutting down the fossil fuel industry begins. Most people are willing to let the government kick that can down the road, as many Australians still don't consider climate change an urgent issue that personally affects them or even understand the reality of the threats we face as a nation – not in some distant future, but right now. Today.

Despite deteriorating conditions all over the world, Australia is yet to have a serious national conversation about the realistic limits of climate change adaptation. Most people do not realise that the latest UN climate reports say there are "hard limits" to what people can adapt to beyond 2°C of global warming in many parts of the world. Despite our inherent vulnerabilities, Australia still does not have a national climate change adaptation plan, as other parts of the world do. Although the government is in the process of developing a national strategy – essentially, we have a plan to make a plan – the pace of progress has been disastrously slow, given the urgency of the crisis we face. While all states and territories have adaptation plans of some kind, the lack of national leadership has meant that regions have had to go it alone without adequate operational resources. Without national coordination and funding, these plans are little more than words on a page. As anyone still displaced from recent weather extremes will tell you, we don't need more bureaucratic reports, we need on-the-ground action. It makes me wonder – do our leaders understand that it might not be possible to adapt to high levels of global warming in Australia? Why is there such a huge gap between the scientific reality of climate change and ordinary people's perception of the crisis we are facing? What on Earth is going on?

Before going much further, it's important we start with a shared understanding of the scientific reality we are now facing. As someone who has been studying, researching and teaching climate science for close to thirty years, I want to emphasise that what follows is not opinion or speculation. It is a summary of reports written by world's leading scientists, policy analysts and researchers. Yes, there are uncertainties in the science – trying to understand the interactions of the Earth's processes is as complex as trying to understand the functioning of the human body – but there is plenty of solid evidence to allow us to move forward and avoid as much risk as possible.

Like medical doctors in an emergency room, climate scientists are recommending things to do to stabilise and monitor the situation – not in 2030 or 2050, but right now. While the aim is always to minimise the risk of harm, it can still be hard to predict the exact outcome. It's the reason why scientists provide a range of future scenarios, multiple lines of evidence and best estimates based on probabilities to back up our recommendations. Just as we do when addressing a serious medical condition, we need to move forward with the clear information we have and adjust our treatment plan as new developments come to light. Right now, we are still doing harm.

In 2022 the IPCC warned: "The cumulative scientific evidence is unequivocal: climate change is a threat to human well-being and planetary health. Any further delay in concerted anticipatory global action on adaptation and mitigation will miss a brief and rapidly closing window of opportunity to secure a liveable and sustainable future for all." Global emissions need to fall 43 per cent by 2030 to have a chance of limiting warming to 1.5°C by 2100, and drop 28 per cent to keep the 2°C option alive. As you'll see, the implications of both these outcomes are disastrous for Australia and the rest of the world.

I understand that the climate debate is often technical and overwhelming, but don't worry – you don't need to be a scientist to follow along. I'll keep things as clear and straightforward as possible. Please, don't look away. There isn't a moment to waste.

The year 2023 was one for the record books – an *annus horribilis* for the ages. Conditions were so extreme that some scientists are now wondering if we are witnessing the precursors of abrupt climate change. The Earth experienced its hottest year on record, caused by a complex interaction of a naturally occurring El Niño cycle in the Pacific Ocean boosted by human-caused global warming. These conditions saw the global average temperature reach 1.45°C above pre-industrial levels as measured across the six major datasets used to monitor the Earth's temperature. In one of the American records, the global average temperature breached the 1.5°C threshold for the first time. In less than a decade, the outlier year of 2023 will become typical of a 1.5°C world, pushing future extremes into even more dangerous territory. Recent observations have shown scientists that sometimes large changes in the climate can be rapid and unexpected. The system can behave in a non-linear, erratic way that is hard to predict. According to the World Meteorological Organization, the previous global temperature record of 1.29°C, set in 2016, was broken in 2023 by a "huge margin," stunning the climate science community. As usual, Secretary-General António Guterres didn't mince his words, saying: "Humanity's actions are scorching the Earth. 2023 was a mere preview of the catastrophic future that awaits if we don't act now."

He's not wrong. 2023 was the first year where temperatures on land exceeded 2°C above pre-industrial levels, with extraordinary heat records broken throughout the world. Seventy-seven countries – together accounting for close to a third of the world's population – reported their warmest year on record, particularly in the Northern Hemisphere, where land areas were 3.8°C warmer than normal. Between June and December, every month set new records, with July, August and September the hottest months ever recorded by international monitoring stations. Australia sweltered through its hottest winter since records began in 1910, with temperatures running 1.53°C above the historical average, contributing to the nation's driest three-month period on record from August to October. Global temperatures in September were an extraordinary 1.75°C warmer than pre-industrial levels – the warmest anomaly of any month in recorded history. The September record sent a shock wave through the research community, with American climate scientist Zeke Hausfather calling the observations "absolutely gobsmackingly bananas."

Perhaps the consequences of 2023's relentless heat were clearest in Canada, which suffered through its own "Black Summer," with wildfires ravaging the nation like no other year in its history. Around 15 million hectares went up in flames – an area over twice the size of Tasmania – more than double the previous national record, set in 1989. As forest ecologist Werner Kurz explained: "2023 was basically a year that was so far off the scale in terms of emissions that we will have to completely revise the way we draw our graphs because the y-axis has to basically be doubled in scale."

Horrendous wildfire conditions unleashed a plume of toxic smoke stretching more than 3000 kilometres across North America. The haze turned the skies above New York an apocalyptic orange, choking residents with air pollution worse than New Delhi's. The vast size of the blazes saw smoke drift further afield, smudging the skies over Europe. People in eastern Canada and the north-eastern United States were forced to wear masks to protect themselves from dangerously polluted air, as hospitalisations for respiratory conditions such as asthma spiked. The extreme wildfire season of 2023 generated Canada's highest carbon emissions ever recorded by the nation by a long shot, releasing an enormous pulse of 2.4 billion tonnes of carbon into the atmosphere. That's three and a half times the annual emissions of the entire Canadian economy. Although the freezing temperature and snow eventually smothered the flames, some outbreaks continue to smoulder underground in peatlands as "zombie" fires, potentially rising from the dead in 2024 when the summer heat returns. It's a grave omen of things to come.

If the extreme heat on land wasn't bad enough, 2023 also set a record for the heat content of the world's oceans. While most people know that greenhouse gases trap extra heat in the Earth's atmosphere, not all of us are aware that the vast majority of heat – some 93 per cent – is actually absorbed by the ocean. Around two-thirds accumulates in the upper 700 metres of the sea surface, with additional heat also mixing into the deep layers of the ocean. In 2023 the heat stored in the ocean was around twenty-five times the total energy produced by all human activity in 2021 (the latest year available in the global dataset). This means that the ocean is now absorbing a colossal amount of heat, reducing its cooling effect on the Earth's climate. It's a staggering statistic to take in – excess heat from the burning of fossil fuels is fundamentally destabilising the energy balance that has kept our current climate in place for thousands of years.

Although these numbers are abstract, the real-world consequences are clear. As the ocean continues to warm, even slight changes in temperature alter the density of water masses that drive the circulation of ocean currents and overlying wind patterns, which, in turn, influence local weather conditions. It's why historically relied-on ocean cycles like the El Niño – a natural fluctuation of ocean temperatures in the tropical Pacific Ocean that has the strongest influence on the world's climate from year to year – are starting to break down in a warmer world, making life harder for weather forecasters. A hotter ocean evaporates more readily, providing more fuel for tropical cyclones and other severe storms, amplifying their destructive impacts as they track further south of their normal range, where infrastructure is not designed to cope with cyclonic conditions. Warmer seas spreading out towards the poles also melt ice more easily in Antarctica and the Arctic, accelerating the speed of sea level rise. We'll come back to these ideas later, but for now I just want you to know that a rapidly warming ocean affects just about every part of the climate system. The bad news is that this warming of the ocean will continue and is now considered irreversible on timescales of hundreds to thousands of years.

In the World Meteorological Organization's analysis of 2023, it reported that from April to December the global ocean experienced an extraordinary nine-month streak of record heat. Once again, September stunned scientists, with ocean temperatures reaching the warmest levels ever recorded by monitoring stations for any month of the year. Relentless heat meant that virtually all the global ocean registered heatwave conditions at some stage during 2023. Unsurprisingly, the impact on marine ecosystems was brutal. Coral bleaching is one of the most obvious visual signs of climate change because reef ecosystems are very sensitive to temperature changes. The reefs of North America, Central America and the Caribbean were particularly hard hit, with experts referring to the conditions as an "unprecedented" mass bleaching event that affected the entire region.

As the Northern Hemisphere summer peaked, ocean temperatures around the Florida Keys in the United States reached a phenomenal 38.4°C, a level commonly found in a hot bath. Record heat triggered severe coral bleaching in a region that has already seen 90 per cent of coral cover disappear since the 1970s. Describing the condition of the Cheeca Rocks reef, US marine scientist Ian Enochs said: "I was not prepared for what I saw there. Every single coral I saw was affected – either bleached or severely paled. It is pretty hard to wrap your brain around. It's a lot to handle." He reported that soft coral species in what was once one of the healthiest and most vibrant parts of the Florida Keys were simply flaking apart and disintegrating in the hot water. The mass bleaching was so severe that some sites in Florida recorded a complete loss of corals. It was a stark reminder of what the scientific community has been warning for years: 70 to 90 per cent of the world's coral reefs will be destroyed with 1.5°C of global warming, with only 1 per cent remaining once 2°C is reached. Our predictions of mass ecosystem collapse are now coming to pass, yet few people grasp the gravity of the crisis.

As expected, the consequences of the Earth's hottest year in at least 125,000 years had diabolical implications for frozen parts of the planet. The exceptional heat of 2023 saw dramatic declines in ice coverage at both poles, the stand-out being sea ice around western Antarctica failing to form during the height of the Southern Hemisphere winter. In June, an area almost the size of Western Australia was found to be "missing" from most sectors of

Antarctica. Sea ice extent was at record lows for eight months of the year, with extreme conditions persisting until well into the spring, making 2023 the third record-breaking year in a row. For those of you interested in statistics, net daily sea ice anomalies observed around Antarctica during 2023 were a horrifying seven standard deviations below the historical average. The steep drop in Antarctic sea ice has alarmed scientists, with the sheer magnitude of the decline suggesting that something very unusual is happening. Some believe that the record high ocean temperature is storing excess heat below the surface that has been accelerating the melting around Antarctica's icy coastlines since 2016. Others fear that recent lows may be the beginning of an "abrupt critical transition" that heralds a regime shift that will result in the inevitable disintegration of the west Antarctic ice sheet.

If the rapid destabilisation of the world's ice sheets doesn't worry you, it should. The problem is that melting ice not only increases sea level in the short term on timescales of years and decades but also unlocks long-term changes that will continue to unfold over hundreds of years into the future. A disturbing number of new studies based on multiple lines of evidence show that if the Earth warms by 2°C, then irreversible melting will be triggered across nearly all of Greenland and much of West Antarctica, committing the planet to between 12 and 20 metres of sea level rise over coming centuries, even if air temperature decreases later. In November 2023, ahead of UN climate talks, a collective of experts on ice sheets known as the International Cryosphere Climate Initiative released a report, warning: "If global leaders cause temperatures to reach this point [2°C above pre-industrial levels] through continued fossil emissions, they are committing the planet to extensive coastal loss and damage well beyond limits of feasible adaptation."

The report then goes on to say that if atmospheric carbon dioxide concentrations continue at today's pace – which has not declined despite Paris Agreement pledges, and in fact has now reached a record high of 418 parts per million – global temperatures will reach at least 3°C above pre-industrial levels by the end of this century. Once 3°C is passed, ice loss from Greenland and West Antarctica may become extremely rapid, with a 3-metre rise in global sea level possible early in the 2100s, wiping out entire low-lying nations and coastlines; with 5 metres breached by 2200; and up to 15 metres in sea level rise possible by 2300. Obviously these changes won't happen overnight. But if such high levels of global warming are reached, they will initiate catastrophic increases in sea level that will be impossible to reverse, altering coastlines and human societies forever.

In the shorter term, the build-up of historical emissions means that there is already a certain amount of sea level rise locked in. The IPCC's *Sixth Assessment Report* states that sea level is expected to rise an additional 10 to 25 centimetres by 2050 regardless of whether emissions are reduced or not. But beyond 2050, further increases depend on the level of greenhouse gases we emit. That is, we still have choices to make about how bad we let things get. In the most fossil-fuel-intensive scenario, global sea level is esti- mated to increase by up to 1 metre by 2100. But given how difficult it is to monitor and model complex ice sheet disintegration processes in remote polar regions, the IPCC suggests factoring in an additional 1-metre rise on top of the likely projected range. So in a scenario where emissions remain high and ice sheets melt faster than represented in the models, global sea level could rise by up to 2 metres by the end of this century.

The 2023 report released by the International Cryosphere Climate Initi- ative, which takes in the latest research published since the IPCC literature cut-off date of January 2021, shows trends are tracking the IPCC's worst-case scenario, making its assessment even more chilling. Yet again, the scientists explain that the possibility of avoiding this monumental scale of coastal inundation will be determined by political decisions being made right now. Their exasperation is clear in the uncharacteristically blunt language of their report: "this insanity cannot and must not continue ... the melting point of ice pays no attention to rhetoric, only to our actions."

The scientific community is clearly at its wits' end; 2023 was a terrifying glimpse into a future we have all been trying to avoid. Reflecting on the extraordinary year, the director of NASA's climate science and modelling program, Gavin Schmidt, said: "It's humbling, and a bit worrying, to admit that no year has confounded climate scientists' predictive capabilities more than 2023 has." He goes on to say that if global temperatures do not stabilise by August 2024, when the El Niño should have dissipated, then "the world will be in uncharted territory." It is clear that climate change is playing out more ferociously and unpredictably than many scientists feared. In the back of our minds we are also privately grappling with this inescapable truth: every year from hereon in will be one of the hottest years on record, and 2023 will end up being one of the coldest years this century.

FIRE, FLOOD AND DROUGHT

In February 2024 I attended the annual conference of the Australian Meteorological and Oceanographic Society – the peak group for scientists working in all branches of weather and climate research. Over the past decade, the mood of our gatherings has become increasingly sombre. Some presenters have taken to apologising in advance for their confronting results, with some attempting to soften the blow by including funny animated gifs or photos of soothing sunsets to comfort the audience. It's not hard to understand why. This year we had a plenary address by a distinguished IPCC veteran. The speaker began by saying that the world has "Buckley's chance" of achieving the 1.5°C target, and even 2°C is going to be a stretch. If emissions continue at the current rate, the 1.5°C threshold could be breached as soon as 2028. Forget the critical decade, what happens every single month during the next handful of years is crucial in determining how quickly we drain the remaining carbon budget needed to achieve the temperature goals of the Paris Agreement. People who have been working in the field for decades are no longer sugarcoating the bad news – they want us to feel an appropriate level of alarm and outrage so we can get on with the job of doing something about the terrible situation we find ourselves in. We need you to stare into the abyss with us and not turn away.

Even a cursory look at the latest figures released by the United Nations Environment Programme (UNEP) shows that the situation we are facing is extremely serious. The 2023 *Emissions Gap Report* – subtitled "Broken Record: Temperatures hit new highs, yet world fails to cut emissions (again)" – explains that, even in the most optimistic scenario, the chance of limiting global warming to 1.5°C is just 14 per cent, with various scenarios indicating a 90 per cent probability of warming between 2°C and 3°C by the end of the century. If currently implemented policies are continued with no increase in ambition, there is a 90 per cent chance that the Earth will warm between 2.3°C and 4.5°C, with a best estimate of 3.5°C. Despite all the political rhetoric you might have heard in the news, the scientific reality is that the planet is still on track for catastrophic levels of warming. Even if nations make good on their net zero promises – which is a big "if" because right now many nations' pledges have no finance, weak implementation or limited political ambition, so are effectively empty promises – there is a 90 per cent chance that we are still on track for 2.4°C of global warming under this best-case scenario, which will lock in centuries of irreversible changes to the climate system.

I know these numbers are hard for most people to absorb, so perhaps the best way to grasp the reality of climate change in Australia is to consider the impacts we've already witnessed so far with 1.2°C of global warming. As the driest inhabited continent on the planet, Australia is particularly vulnerable to climate change. Our nation is a huge island surrounded by the Pacific, Indian and Southern oceans, resulting in dramatic swings in our weather. You can think of Australia's background climate as essentially a tug of war between warm tropical influences from the north and cool temperate systems from the south. The weather experienced from season to season is driven by differences between the rate of warming of the hot land and surrounding cool ocean. These contrasting temperature and air pressure gradients set the scene for the complex interaction of atmospheric and ocean cycles that drive Australia's highly variable climate. These factors make weather and climate forecasting very difficult in our part of the world – there are a lot of complicated dynamic processes that are hard to represent with the mathematical equations used to drive climate models. It's a bit like trying to reduce the functioning of each part of the human body down to lines of computer code. It's very, very difficult – if not impossible – to capture the full complexity of behaviour. This is especially true in a rapidly warming climate which is now altering historical weather patterns, making them more erratic and harder to predict. Nonetheless, there is still a huge amount we can say about the operation of the Earth's climate, with advances in computing technology and our fundamental understanding of the science rapidly improving our models with every passing year.

To add to the complexity, the landmass of Australia stretches from the tropics in places like Far North Queensland to the temperate mid-latitudes of Tasmania, generating an enormous range of climate zones that sustain rainforests, coral reefs, deserts and alpine environments. This is why we can have tropical cyclone conditions, extreme heatwaves and bushfires happening at the same time across the country, challenging emergency services. Australia is also the flattest continent on Earth, meaning that weather systems can travel vast distances without being tripped up by rugged terrain, unleashing destruction over large areas. The mountain chain that moderates the weather and climate of Australia's east coast is the Great Dividing Range; a feature that stretches 3500 kilometres from the northern tip of Queensland, running the entire length of the east coast before disappearing into the central plains of western Victoria. When weather systems collide with mountains, there can be rapid uplift of air masses, leading to atmospheric instability that can result in severe thunderstorms and torrential downpours that trigger flash flooding and destructive winds. But essentially, away from the eastern seaboard, Australia is mostly a flat, dry desert with wet coastal fringes that house our capital cities. Today, close to 90 per cent of Australia's population lives within 50 kilometres of the coast, and that number is increasing, leaving us particularly exposed to the threat of sea level rise.

When thinking about the effects of a warming planet, it is important to understand that the world does not heat up uniformly. Because of the presence of oceans, mountains, glaciers and forests, different areas of the Earth warm at different rates. The geographical characteristics of Australia mean that the continent is warming faster than the global average – temperatures have risen by 1.5°C since 1910, compared with the 1.2°C global increase since pre-industrial times. Given that around three-quarters of Australia is already classified as arid or semi-arid – with half of the country receiving less than 300 millimetres of rainfall each year – further warming threatens to make life on an already very dry continent even harder. We are very vulnerable to intense swings in rainfall that cause droughts and floods, relentless heat, and the risk of permanent inundation of low-lying areas from rising seas.

As global warming continues, Australia's climate is fast becoming more extreme and unpredictable, edging us closer towards breaching thresholds that will make it very difficult, if not impossible, to adapt to. This is especially the case when there are simultaneous disasters unfolding in different regions, or a rapid succession of back-to-back disasters that undermine the ability of communities to recover. If there is not enough time between destructive events, the damage begins to compound. We see the continued degradation of our natural environment and the weakening of social resilience that will eventually lead to the permanent displacement of people from their homes and ongoing impacts on our economy.

We don't need to use our imagination to picture what this scenario looks like. The Black Summer bushfires of 2019–20 and the 2022 east coast floods highlighted Australia's lack of preparedness. Our emergency services were alarmingly under-resourced and stretched thin across vast areas, which left many local communities to fend for themselves. During the catastrophic flooding of the town of Lismore, in northern New South Wales, in 2022, we witnessed extraordinary scenes of locals rescuing each other from rooftops in their boats, jet skis and kayaks when the handful of State Emergency Service crews were overwhelmed by the needs of 45,000 residents. People in rural areas set off in their boats with cordless angle grinders to cut people out of the roof cavities of their homes where they'd been forced to retreat. The situation was so bad that the army had to be called in, but it did not show up until a full five days later, leaving the terrified community feeling abandoned.

Over two years on, people from Lismore are still displaced from their homes and uncertain about how to move forward. Do they plan to relocate the town and pray that another once-in-a-century flood won't happen again, or is the writing already on the wall and it is time simply to abandon ship? Although fifty-one countries around the world have national climate adaptation plans, Australia is not one of them. This colossal failure of government has left many people wondering what is going to happen to our communities as climate change continues to worsen. How many disasters does it take to wake people up to the fact that Australia's climate is becoming more extreme, with today's destruction set to be dwarfed by things to come? Do people realise that adapting to climate change won't be possible in some parts of the country? Exactly how much do we need to lose before our political leaders decide to take this seriously?

*

Recent extremes are a good way of picturing what climate change might look like in the future. The Black Summer bushfires started in 2019, Australia's hottest and driest year on record. The fires tore through extensive areas of vegetation that rarely burn, including fire-sensitive rainforests and alpine ecosystems that once acted as a natural barrier to the spread of wildfires. An area containing more than one-third of all Australian plant species burnt, including 44 per cent of Australia's threatened plant species. Usually wet, subtropical rainforests ignited during winter, incinerating over half of the country's ancient Gondwana-era rainforests in a single bushfire season. Although eastern Australia's eucalyptus forests are among the most fire-prone in the world, typically only 2 per cent burn during extreme fire seasons. But in 2019–20, almost a quarter of Australia's temperate forests went up in smoke, setting a new global record for the sheer scale of the blazes. Not an easy feat for a nation with a formidable bushfire history.

The marked increase in the area burnt has led scientists to develop a new category of wildfires: "megafires." This describes an individual wildfire or wildfire complex that engulfs more than 1 million hectares, or an area almost the size of the greater Sydney region. Australia's record-setting megafires burnt a phenomenal 24 million hectares, roughly the size of the state of Victoria. The fires released over 715 million tonnes of carbon dioxide: more than all the emissions Australia releases in an entire year. It's an important idea I'll come back to later, but for now I just want to flag that climate policy that is over-reliant on offsetting carbon emissions in a bushfire-prone country like Australia is not a very smart idea.

The environmental and societal impacts of the Black Summer bushfires were enormous, culminating in economic losses of over $10 billion and the destruction of 6000 buildings, including over 3000 homes. Alongside thirty-three direct human deaths from the fires and 429 more from smoke inhalation, 3 billion animals were killed or displaced by the immense scale of habitat destruction as fire ripped through globally significant biodiversity hotspots across the country. The impacts on wildlife were profound. The koala, our most emblematic species, lost so much of its habitat that it now faces extinction in New South Wales as early as 2050. The situation following the fires was so bad that in February 2022, koalas in eastern Australia were officially added to the endangered species list – something I never imagined I'd witness in my lifetime.

While the impacts of the catastrophic fire season on our ecosystems and society are still being studied, what worries me most is what Australia's Black Summer says about things yet to come. In the aftermath, researchers analysed the conditions observed during the 2019–20 fire season and concluded that "under a scenario where emissions continue to grow, such a year would be average by 2040 and exceptionally cool by 2060." It's a sickening thought – the most extreme bushfire conditions in today's climate will be considered average in less than twenty years. This means a future where there is less time for ecosystems to recover between severe burns, leading to a degradation and eventual loss of tracts of native forests and all the creatures they contain. Although it is hard to conceive what repeated, high-intensity wildfires will do to our landscapes and communities in a future of unmitigated warming, it is something we need to plan for as the world is currently failing dismally at reducing emissions.

In recent years, climate scientists have researched what global warming will mean for every part in the world, including Australia. The IPCC provides extensive regional information detailing how everything from temperature, rainfall, sea level and ecosystems are projected to change under different emissions scenarios. If the world's collective Paris Agreement pledges for 2030 are fully implemented in tangible policies, our emissions trajectory is approximately in line with the IPCC's "middle of the road" intermediate-emissions pathway (the IPCC refers to this as Shared Socioeconomic Pathway SSP2–4.5). Under this optimistic scenario, Australia's average temperature increases between 2.2°C and 4.1°C by 2100, with a central estimate of 3.1°C. That's twice the warming Australia has already experienced.

If we consider a future of resurgent nationalism and the continued investment in fossil fuels represented in the IPCC's high-emissions scenario of SSP3–7.0, Australia warms between 3.2°C and 5.6°C by 2100, with a best estimate of 4.4°C. I consider this the most plausible scenario based on currently implemented policies – not on well-intentioned promises – and uncertainties in climate modelling such as carbon cycle feedbacks, which are expected to amplify warming, and the higher sensitivity of temperature to increases in carbon dioxide than previously estimated.

The IPCC also considers a worst-case, very-high-emissions scenario, known as SSP5–8.5, which is based on unconstrained economic growth based on abundant fossil fuel use and a failure to substantially reduce emissions before 2050. While some experts believe that major improvements in renewable energy technology and the overly high sensitivity of some climate models make this an unlikely outcome, it is still considered possible to reach very high levels of warming even under lower-emission pathways. But just for the sake of completeness, if the world finds itself in the nightmarish future of underestimated climate feedback loops and complete policy failure represented by the IPCC's SSP5–8.5 scenario, Australia's average temperature warms 4.0°C to 7.0°C by 2100, with a central estimate of 5.3°C. And just to be clear – we are just talking about average, background warming here, not maximum temperatures, which often increase twice as fast as average conditions.

I know I've given you a lot of numbers to digest, so let me try to explain what such high levels of global warming mean for life in Australia. Climate modelling studies have shown that an increase of "just" 2°C in the Earth's average temperature will lead to days above 50°C in Sydney and Melbourne as early as the 2040s, and they would become a regular feature of the Australian summer at 3°C of global warming. Our two largest cities, which house 40 per cent of the nation's population, will be exposed to life-threatening levels of heat under this new climate. This would limit the amount of time that we can spend outside without subjecting ourselves to heat capable of killing us every summer. It would also mean facing the risk of longer and hotter bushfire seasons spreading further and further into areas not previously considered fire-prone. Higher temperatures also result in a thirstier atmosphere capable of sucking moisture out of waterways, soil and vegetation, making droughts longer and more intense than they were in the past. Under such high levels of warming, inland parts of the country will start to desertify as intense heat and aridity take hold.

The effect of increasing aridity will be clearest on the edges of current rainfall zones. This is already happening in south-western Australia, where there has been a 20 per cent decline in winter rainfall since 1970, leading to water shortages, forest dieback and agricultural losses. The dry margin of Australia's grain belts will continue to dry out, transforming previously productive farmland into areas only capable of sustaining grazing. And even then, heat stress will threaten the welfare of livestock and farmers' ability to feed their herds, especially during droughts. The Australian Academy of Science notes that climate change has already affected our croplands, with wheat and barley recording a 27 per cent slash in yields in recent decades. Sustained water stress will threaten Australia's food security and limit our capacity to contribute to the international trade of wheat, meat and dairy products. More and more farmers will walk off the land, as previously productive agricultural areas turn into dust bowls.

Such major shifts in our temperature and rainfall patterns will gradually alter the fundamental climate of our nation, moving us beyond conditions we've historically relied on. In 2015, the CSIRO and Bureau of Meteorology released an online tool that made use of "climate analogues" to help explain what future climate change might be like relative to other places in Australia. In response to a specific question like "What will the future climate of Sydney be like?" a database of climate model simulations statistically identifies locations where the current climate is similar to the projected future climate of a place of interest. So for example, under a high-emissions scenario Sydney takes on the subtropical climate of Brisbane, while Melbourne and Hobart transition to the aridity currently experienced in the South Australian capital.

By the end of the century, Darwin will not resemble any part of modern-day Australia: an entirely new climate will have formed, as monsoonal conditions experienced in neighbouring South-East Asia spread into northern Australia. This extension of the tropics deeper into our north means we will see tropical cyclones drifting into areas on the southern edge of current cyclone zones, into places such as South-East Queensland and northern New South Wales, where infrastructure is not built to withstand cyclonic conditions. These areas currently house close to 4 million people, leaving the third-largest population centre in the country at risk of extensive damage from severe storms supercharged by warmer seas and a wetter atmosphere in a hotter world.

The effects of sea level rise on our coastal nation are huge. The bad news is that regardless of whether the world takes a low- or high-emissions pathway, several metres of sea level rise are now locked in over the coming centuries. This is because of the slow response of the oceans and ice sheets to historical emissions as the Earth finds its new equilibrium. But the good news is that the choices we make now will determine how quickly these impacts are realised. For context, over a quarter of a billion people currently live on land less than 2 metres above mean sea level, so are already in harm's way. As a rule of thumb, coastal experts estimate that every 10 centimetres of sea level rise triples the frequency of a given coastal flood, with every 1-metre rise in sea level resulting in a 100-metre retreat of the coastline. As you might imagine, trying to relocate a huge proportion of the human population displaced by eroding coastlines and repeated inundation from more frequent storm surges is likely to be impossible, particularly in a geo-politically fragile world.

What global sea level rise translates to in a specific region of the world depends on several factors, including the shape of the sea floor, tidal processes and the presence or absence of coastal protection, such as natural wetlands and dunes or infrastructure like sea walls and levees. Depending on these features, local sea level changes can be as much as 30 per cent lower or higher than the global mean. Since the nineteenth century, average sea level has increased globally by more than 20 centimetres and is now rising at an accelerating rate. In Australia sea level is rising at or above the global average, with increases of around 5 millimetres per year in northern Australia and 3 millimetres per year along the south-east coast. As warming continues, there will be a higher risk of coastal inundation during high tides and storm surges associated with severe weather systems such as east-coast lows and tropical cyclones.

Under the IPCC's "middle of the road" scenario (SSP2–4.5), the sea level along much of eastern Australia rises between 30 to 80 centimetres above recent levels by the end of the century, with an average rise of 60 centimetres. If the fossil fuel industry continues to expand, the SSP3–7.0 pathway sees an extra 10 centimetres added to the problem – close to a 1-metre rise along Australia's east coast over the next seventy years. Such a substantial increase will have major consequences for our environment and society. Even under the IPCC's intermediate-emissions scenario, Australia's sandy shorelines are projected to retreat by around 110 metres in eastern Australia and up to 90 metres on the south coast by the end of the century. Under the worst scenario, we see 100 metres of our coast disappear around the country, with retreats as high as 220 metres in northern Australia and 170 metres in eastern Australia, where most Australians live. Aside from the loss of large chunks of our coastline, severe storm surges are expected to further exacerbate coastal erosion, meaning we will witness many of our beloved beaches degrade and disappear within the lifetime of children alive today.

Currently over half the Australian coastline is vulnerable to erosion from rising sea levels, with 80 per cent of the Victorian coast and close to two-thirds of the Queensland coast considered at risk. Despite the nation's high exposure to sea level rise, up-to-date figures are surprisingly hard to come by. In 2011, an analysis published by the federal Department of Climate

Change and Energy Efficiency estimated that more than $226 billion in commercial, industrial, road, rail and residential assets around Australia are potentially exposed to inundation and erosion hazards from 1.1 metres of sea level rise. It reported that between 26,000 and 33,000 kilometres of roads are at risk from the combined impacts of coastal inundation and shoreline recession, valued at up to $60 billion in 2008 dollars. The analysis also identified up to 274,000 residential buildings that are exposed to a sea level rise of just over a metre, with New South Wales and Queensland residents most vulnerable to inundation and erosion. A recent study by the Victorian government, which also considered a 1.1-metre sea level rise scenario, found that a quarter of Melbourne's Port Phillip council area – which includes St Kilda, Albert Park and Southbank – could face inundation during a one-in-100-year storm tide by the end of the century.

To get a vivid sense of what changes to Australia's coastline might look like with sea level rise, Coastal Risk Australia has developed an online mapping tool that uses Google Earth and the latest IPCC estimates in conjunction with national high-tide data and information about the shape of our coastline. If you choose the IPCC's high-end scenario of a possible 2-metre sea level rise by 2100 under a fossil-fuel-intensive pathway that accounts for ice sheet instabilities, the sheer scale of inundation is staggering. As someone who lives on the coast and visits the beach most days, I found looking at these images very disturbing. The maps show that significant areas of Australia's major cities and crucial infrastructure could be underwater if not by the end of this century, then in years to come, even if warming is limited to below 2°C.

With a 2-metre rise in global sea level, airports located near the coast in Sydney, Brisbane, Adelaide, Hobart and Cairns would largely be underwater at high tide. Many of Australia's most densely populated areas could be at risk of becoming uninhabitable, or subject to an increased risk of destructive storm surges as the sea rises. For example, in Sydney, Circular Quay, the Royal Botanic Gardens, Pyrmont and the Barangaroo precincts are all inundated. Further north, low-lying parts of Newcastle, Port Macquarie, Ballina and Byron Bay will be among the most heavily hit, as rising seas surge into coastal areas. In Victoria, Melbourne's southern suburbs of Port Melbourne, St Kilda and Docklands, as well as the CBD, are the worst affected. Vast areas of the Gold Coast, Cairns and Noosa in Queensland, and the WACA ground and Cottesloe Beach in Perth, are also engulfed in this 2-metre sea level rise scenario. In the Northern Territory, huge swathes of the city of Darwin are inundated, including the proposed site of the Middle Arm petrochemical precinct designed to exploit vast amounts of gas from the Beetaloo Basin and offshore fields. In this future, maps of the world – not just Australia – will have to be redrawn as the sea level rises, altering our coastlines in response to a drastically altered climate.

*

As you can see, the scientific reality of a rapidly warming world is very confronting, especially in Australia. As the years since our Black Summer have shown us, the prospect of a future of dealing with back-to-back disasters across the country every year is ultimately going to be impossible for our ecosystems and communities to adapt to. The conditions experienced in 2023 have many experts worried that we may have breached regional and global tipping points that will unleash a cascade of changes that will be with us for thousands of years. The problem is that we will only know if we have definitively crossed critical thresholds for planetary stability in hindsight, so we have to move forward armed with the best available science while we still can to minimise the damage.

The latest research shows that several tipping points, such as the disintegration of the Greenland and West Antarctic ice sheets, may be triggered within the Paris Agreement range of 1.5°C to 2°C of global warming. This means it is possible that the Earth will experience major transformations even if we manage to achieve the goals of the Paris Agreement. In a 2024 report released by the CSIRO on the risks of tipping points to Australia, the authors warn: "The effects of tipping points on the global climate are generally not currently accounted for in projections based on climate models.

This means that effects of tipping points are also not included in national climate projections and impact assessments for Australia and may represent significant risks on top of the changes that are generally included." The report suggests the need to plan for "low likelihood high impact" scenarios that include climate tipping points. For example, the construction of new critical infrastructure should incorporate global sea level rise scenarios of around 2 metres by 2100. Given that we are a highly coastal nation, the adaption challenge of planned relocation and retreat will be enormous. While it may be possible to protect vulnerable areas with sea walls or the restoration of natural dunes for a while, these are only band-aid solutions that will not stem the rising tide for long. Hard decisions will need to be made by local councils around the country about how and when they plan to withdraw residents from high-risk areas. As the Australian Academy of Science notes: "Under high levels of warming and sea level rise, retreat is likely to be the only feasible long-term strategy."

Scientists know the situation is very bad, but we also know exactly what we need to do to stabilise the Earth's temperature and avoid triggering a domino effect of impacts in other components of the climate system. If we don't put the brakes on industrial emissions immediately, children alive today will inherit this nightmarish future. It makes me wonder if people in decades to come will look back at the world's collective failure to shut down the fossil fuel industry in time and see it for what it really is: an intergenerational crime against humanity.

Political recognition of the threat of climate change began over thirty years ago, following the Intergovernmental Panel on Climate Change's (IPCC) *First Assessment Report* in 1990. Acknowledgement that human activity can dangerously influence the stability of the Earth's climate can be traced back to the United Nations Conference on Environment and Development held in Rio de Janeiro, Brazil, in 1992. Often referred to as the Rio Earth Summit, the landmark meeting resulted in the first international climate treaty, the United Nations Framework Convention on Climate Change (UNFCCC), which came into force in 1994 with the goal of avoiding "dangerous human interference with the climate system." Since 1995, there have been yearly meetings of the Conference of the Parties (COP) signed up to the UNFCCC to assess global progress. After decades of diplomatic discussion and failure to reduce greenhouse gas emissions, it wasn't until COP21 in 2015 that 196 nations – including Australia – signed the historic Paris Agreement. The goals of the legally non-binding pact are to keep global warming well below 2°C and as close as possible to 1.5°C above pre-industrial levels; for global emissions to peak as soon as possible; and for the world to reach net zero emissions in the second half of this century.

In December 2023 world leaders gathered in Dubai, in the United Arab Emirates, to attend COP28. Presiding over proceedings was Sultan Al Jaber, CEO of the Abu Dhabi National Oil Company, a corporation planning a huge expansion of its oil and gas production. A conflict of interest if there ever was one. The official purpose of COP28 was to provide the first global stocktake of how each nation is tracking to meet its pledges – technically known as Nationally Determined Contributions – needed to achieve the world's collective goal of limiting warming to below 2°C. But in reality the political goal of COP28 was for world leaders to provide clear guidance on the future of the fossil fuel industry. It might surprise you to learn that in the entire history of the UNFCCC process, resolutions arising from these meetings have never explicitly mandated or even mentioned the end of fossil fuel production, despite it being the root cause of global warming. Denial of this most fundamental truth encapsulates just how disgraceful the political response to climate change has been over the past thirty years.

Although COP28 delivered text that contained the words "fossil fuels" for the first time in the UNFCCC's history, the phrasing around whether to "phase out," "phase down" or "transition away" from them was fiercely debated. The outcome really mattered, as weak language would allow room for the continued exploitation of coal, oil and gas reserves, locking in irreversible changes to the Earth's climate for centuries to come. In the end, the most watered-down wording of "transitioning away from fossil fuels" was the only consensus that could be achieved, to the immense disappointment of small island nations and many in the scientific community. Some of the most insidious wins for the fossil fuel industry were the inclusion of clauses calling for more CCS technology to justify "business-as-usual," and the use of "transitional fuels" to legitimise the burning of natural gas (methane) on the basis that it is relatively less polluting than coal. These efforts serve to divert investment away from renewable energy, reinforcing the reliance on fossil fuel infrastructure for decades to come.

The other major victory for the fossil fuel industry was the call for the "phase down of unabated coal power," allowing the use of the largest contributor to global emissions to continue. This means that the use of *"abated"* coal, which refers to the hypothetical capture and storage of some proportion of greenhouse gases released from the burning of fossil carbon, is still very much on the table. To be frank, this is the language of greenwashing, championed by major fossil fuel producers, including Saudi Arabia, Australia and Russia, to allow polluting industries to keep on polluting. The insertion of loopholes that rely on non-existent technology to preserve the status quo is disastrous for the climate, as it delays facing the need to stop burning fossil fuels as soon as humanly possible. The scientific reality is simple: without a clear, legally binding deadline to phase out the use of coal, oil and gas, we will be unable to avert the worst aspects of climate change and we increase the risk of making some parts of the world – including areas of Australia – uninhabitable.

*

A friend I admire deeply is in the difficult position of representing the Australian government's conflicted stance in global climate negotiations. Given rising geopolitical tensions in the world right now, their one-word assessment of the state of current climate talks was "soul-destroying." There is still a lot of fighting over who is responsible for global warming and how much the most affected nations should be compensated for a problem largely not of their making. Major fossil-fuel-producing nations such as Saudi Arabia, Russia and China are blocking diplomatic progress to increase global ambition to keep the Paris Agreement targets alive, knowing full well that the most vulnerable people in places like Pacific Island nations will be forcibly displaced from their homes even with modest levels of sea level rise. Not their problem, it seems.

According to the UNEP, the global community spent US$21 billion helping developing countries adapt to climate change in 2021. This is even though the UN has estimated that the cost of adaptation for developing nations alone is as high as US$387 billion each year out to 2030. To limit the degree of adaptation needed, between US$4 trillion to US$5 trillion needs to be invested in clean energy each year until 2050 to achieve net zero emissions. While these numbers might sound big, by comparison the International Monetary Fund estimates that in 2022, world governments spent around $7 trillion on fossil fuel subsidies. It's outrageous to realise that the money spent supporting the fossil fuel industry could cover the funds needed by developing nations to adapt to climate change as well as finance the clean energy transition, and still have change left over. And yet the global stocktake text resulting from COP28 notes with "deep regret" that rich nations have still failed to come up with the US$100 billion per year in climate finance promised under the Paris Agreement to help poorer nations adapt. Although a goal has now been set to at least double adaptation finance by 2025, the history of these negotiations tells us that we shouldn't hold our breath.

Right now, adaptation programs are severely underfinanced, leaving vulnerable nations underprepared for the increasing threat of a rapidly warming world. The UNEP reports that the fifty-five most climate-vulnerable economies alone have already experienced losses and damages of more than US$500 billion in the last two decades. These include our neighbours Fiji, Vanuatu and Tuvalu. The implications of not stabilising climate change will result in a national security threat from increased pressure to house displaced people. In 2022, a record high 71.1 million people were internally displaced globally, mostly by conflict and violence, but also by weather-related disasters that generated 31.8 million refugees. Notably, the displacements from extreme weather were mostly triggered by flooding in Pakistan, the Philippines and China following the three-year La Niña event, which also resulted in catastrophic floods in eastern Australia.

Australia is drastically unprepared for an influx of climate change refugees. In the financial year 2022/23, planned overseas migration added 518,000 people to Australia's population – the largest net increase since the Australian Bureau of Statistics began keeping records. Although such a high level of migration has helped keep Australia out of economic recession, a larger population will increase domestic emissions and exacerbate the housing shortage, driving developers into more marginal, high-risk zones like floodplains and bushland areas. Any insurance company will tell you that climate change is already pushing insurance premiums higher. Analysis by the Climate Council shows that over half a million properties are at high risk of being exposed to annual damage costs from extreme weather and climate change that will effectively make them uninsurable by 2030. River flooding poses the biggest risk, particularly in South-East Queensland and northern New South Wales, where pressure for new housing developments is concentrated. It's hard to imagine the pressure our nation will face trying to house more people in a rapidly warming climate that we may not be able to adapt to.

Estimates of the global population at risk of displacement due to sea level rise vary from tens of millions to hundreds of millions of people, depending on future emissions, land elevation and population growth. Some of the latest research shows that if global sea level reaches 1 metre above present levels by the end of the century, at least 410 million people will be at risk of inundation, particularly in major river delta areas of tropical Asia. The latest IPCC figures suggest that the situation may be even worse, with approximately 1 billion people living in low-lying coastal areas and small islands projected to be at risk from sea level rise and storm surges as early as 2050 rather than by the end of the century.

Of course, some regions are more vulnerable than others, with around half of the Pacific's population estimated to live within 10 kilometres of the coast, with at least half of their infrastructure located within 500 metres of the water. The IPCC estimates that with 1.5°C of warming, up to 560,000 people in small island nations are at risk of permanent inundation, with the number increasing to up to 640,000 with 2°C of warming, which could occur as soon as the 2040s.

As the shortfall in global adaptation funding shows, the world is nowhere near being prepared for the scale of population displacement this level of climate disruption would unleash. In light of the scientific reality of the problem, Australia's offer to take in up to 280 climate change refugees from the Pacific Island nation of Tuvalu each year is laughable. And deeply insulting, given that Australia's Department of Industry, Science and Resources stated in December 2023 that "Oil and gas projects remain a big part of Australia's investment landscape," listing plans for 100 fossil fuel developments, mostly gas projects located in Western Australia, operating for decades to come. If all these projects go ahead, they will collectively add 3.2 billion tonnes of emissions to the atmosphere by 2030 – or nearly seven times Australia's total domestic emissions.

*

While some commentators hailed the very mention of fossil fuels in the COP28 decision text as a historic sign of the "beginning of the end" of the most destructive industry on the planet, I shared US climate scientist Peter Kalmus's assessment:

> Words like these – weaselly, unbinding, unquantitative, insincere – serve to distract society from ending fossil fuels. For 30 years, fossil-fuel-industry-influenced rich people have distracted everyone so effectively that they didn't even need to bother mentioning fossil fuels. And fossil-fuelled "business-as-usual" has continued and expanded exponentially all the while. As a society, we so far seem incapable of imagining it ending.

Perhaps this failure of imagination is not too hard to understand. COP28 saw a record number of fossil fuel lobbyists attend the UN climate talks. Close to 2500 delegates representing the interests of corporations such as Shell, British Petroleum and ExxonMobil outnumbered every country delegation aside from the host country of the UAE and Brazil, which will run COP30 in 2025. How can a climate summit imploring world leaders to urgently avert planetary disaster be crawling with fossil fuel lobbyists? It's like the tobacco industry showing up to a lung cancer conference to promote their products. People will look back one day and see COP28 for what is really was: the moment the world turned its back on the 1.5°C target.

The uncomfortable truth is that, despite all the talk, our political leaders are still failing us. Given how long we've known about the problem and how much is at stake, this failure is unforgivable. Polite incrementalism is getting us nowhere fast. We are spinning our wheels. Ultimately our leaders are choosing to protect corporate interests over the very stability of our planet – a decision that can't be undone. In the aftermath of COP28, the Gas Exporting Countries Forum released a joint statement with the Organization of the Petroleum Exporting Countries (OPEC) – which control nearly 80 per cent of the world's oil reserves – offering their "heartfelt congratulations" to the UAE for a "positive outcome" for the fossil fuel industry. The statement emphasised that "oil and gas currently represent 55 per cent of the energy mix and it will maintain a majority share for decades to come" and reiterated that continued investment is essential to meet future demand. Speaking at an oil industry conference in March 2024, Amin Nasser, chief executive of Saudi Aramco, the world's largest oil company, said: "We should abandon the fantasy of phasing out oil and gas, and instead invest in them adequately" to a round of applause from the room.

According to a damning analysis published by the Stockholm Environment Institute and the UNEP, it doesn't look like maintaining global demand for fossil fuels will be much of a problem. The 2023 *Production Gap* report reveals that, despite stated climate pledges, world governments are planning to produce more than double the amount of fossil fuels in 2030 than is consistent with limiting warming to 1.5°C. Collectively, there are plans to increase global coal production until 2030 and oil and gas reserves until at least 2050, despite many major producer countries pledging net zero emissions targets by mid-century. We are rapidly destabilising the Earth's climate, and yet our governments are saying one thing and doing another, pushing the long overdue task of phasing out fossil fuels further and further into the future, saddling our children with an intractable mess.

Although seventeen of the top twenty major fossil-fuel-producing countries, including the United States, Australia, Saudi Arabia and Russia, have promised to achieve net zero emissions, they continue to promote, subsidise and plan for the expansion of fossil fuel production. Analysis by Oil Change International shows that just five countries – the United States, Canada, Australia, Norway and the United Kingdom – are on track to be responsible for over half of all new oil and gas production by 2050. Despite these rich nations having the economic means and a moral responsibility to lead the phase-out of fossil fuels, they are choosing to make the problem worse. These decisions ignore the fact that governments – not the private sector – must play a leading role in setting the direction of fossil fuel production. State-owned entities control half of the world's oil and gas reserves and over half of global coal supplies. What governments do to support polluting industries influences public and private investment, undermining the transition to renewable energy by locking in fossil-fuel-based infrastructure for decades to come. Given that the world must rapidly decarbonise to avert planetary disaster, many projects that are planned or under development are at risk of becoming stranded assets if the global demand for fossil fuels peaks and starts to decline this decade to keep the possibility of achieving net zero by 2050 alive.

*

Despite Labor's grandstanding about its credentials as a climate leader, in March 2023 the federal Minster for Resources, Madeleine King, signalled the continuation of "business as usual" in her Resources Statement to Parliament, saying: "Some fail to acknowledge this, but Australia's coal and gas resources are essential for energy security, stability and reliability both domestically and across the Asia-Pacific and will be needed for decades. From Hanoi to Hyderabad, Seoul to Singapore, families rely on Australia's natural resources to provide the energy security they need." She emphasised that gas is critical to our trading partners' net zero pathways, saying that as nations such as Japan and South Korea pursue their net zero targets, they will continue to require gas for decades. But the truth is that transitioning from coal to gas is like switching from cigarette smoking to vaping – the outcome is still very bad for your health and simply delays the inevitability of having to quit smoking altogether. Given the rapidly deteriorating state of the climate, time is the one thing we no longer have. Every day we delay, we risk breaching critical thresholds needed to maintain planetary stability.

Promoting gas as a necessary transition fuel also fails to consider the fact that renewable energy is now the cheapest form of electricity in the world and is growing exponentially. The International Energy Agency's *Renewables 2023* report details how the world is on track to build enough solar, wind and other renewables over the next five years to power the equivalent of the United States and Canada. It also highlights that onshore wind and solar power are cheaper than both new and existing fossil fuel plants, and that the price of solar panels halved in 2023, driven by China, the world's renewable powerhouse. Here in Australia, while Labor has an ambitious target of 82 per cent renewables in the electricity sector by 2030, they still only accounted for around 36 per cent of electricity generation in 2022. That means the remaining 64 per cent is still being generated by fossil fuels, with coal alone accounting for 47 per cent. Although the development of the renewable energy sector has started to reduce Australia's reliance on coal, declines have been modest, with the share of coal in the electricity generation mix falling from close to 60 per cent in 2021 to around 55 per cent in 2022. Over the same period, gas increased its share of power generation from 7.7 per cent to 8.9 per cent. But according to the Clean Energy Council, Australia's investment in the clean energy industry was $6.7 billion in 2022; that's just over half the amount spent by the government subsidising the fossil fuel industry.

The government's reluctance to move away from polluting industries stems from the fact that Australia is the third-largest fossil fuel exporter. Energy companies reap tens of billions of dollars each year selling coal and liquified national gas (LNG), mainly to the North Asian countries of Japan and South Korea. Right now around 90 per cent of coal and three-quarters of all gas produced in Australia is exported. In 2024, the Climate Change Authority reported that emissions generated from the use of Australia's coal and gas overseas amount to approximately three times those generated domestically, accounting for around 4 per cent of global emissions. But what is odd is that the Authority says: "Customers of Australia's fossil fuel imports will decide when they phase them out and there is a risk that as Australia reduces its fossil fuel exports, other countries may increase their exports to fill the gap." Instead of taking the ethical lead and choosing to limit the supply of products we know are cooking the planet, we are using the drug dealer's defence that if addicts don't buy from us, they'll get their fix somewhere else. Instead of enabling countries hooked on their dirty habit, Australia could instead choose to take them to rehab and help them get clean.

The main culprit is Japan, which gets about 40 per cent of its LNG from Australia, with the head of the Japanese energy giant Inpex, Takayuki Ueda, telling parliamentarians in April 2023 that he feared any restrictions imposed on new gas developments could result in "a direct threat to the rules-based international order essential to the peace, stability and prosperity of the region, if not the world." Although tensions with China are thawing, it is a reminder that energy policy is closely tied to geopolitics, which is becoming increasingly harder to navigate in a war-torn world. Fearmongering aside, the truth is that Japan's net zero plan is heavily reliant on CCS technology, the fantasy get-out-of-jail-free card that threatens to ruin us.

While I don't doubt Labor's intentions are good, and I accept that the challenge the government faces is monumental, unfortunately it is still unwilling to address the root cause of the problem, which is Australia's continuing reliance on fossil fuels. The industry's dominance over our political system is currently blocking the scale and speed of the renewable energy transition. Despite Climate Change Minister Chris Bowen's comments at COP28 that a "phase-out of fossil fuels is Australia's economic opportunity as a renewable energy superpower," we are yet to realise this vision. Despite all the talk, the Climate Change Authority's assessment of progress towards reducing emissions concluded that: "Australia is not yet on track to meet its 2030 targets."

The problem is that many gas producers like Australia are promoting methane as a "transition" fuel to support the shift away from coal, while also using the need to protect their export markets as a reason why we can't reduce our reliance on fossil fuels any time soon. Although the burning of natural gas for electricity is generally considered less emissions-intensive than coal, unintentional "fugitive" emissions released during the extraction, processing, storage and transportation processes can offset the benefit of its lower carbon intensity. The other major issue is that even though methane breaks down more quickly in the atmosphere, it is around eighty-six times more potent than carbon dioxide, measured over a twenty-year period, meaning that the burning of gas will exacerbate global warming in the near term. So while methane may be sold to the public as the lesser of two evils, rest assured that it is still a highly destructive greenhouse gas.

Given that Australia is one of the world's top two exporters of LNG and coal, fossil fuel lobbyists exert a very strong influence on our political debate, diplomacy, economic strategy and policy development both domestically and internationally. Trying to pivot away from fossil fuels has proved to be political poison in Australia, time and time again, despite the nation being highly vulnerable to the impacts of a changing climate. As our next federal election approaches, we need to ask ourselves if we are going to be the ones to remove the social licence for the continued exploitation of Australia's coal, oil and gas reserves – whether for domestic or international consumption – to avoid destabilising the Earth's climate. Is Australia honestly doing enough to secure a liveable future? Or will corporate bullies stop us from doing what really needs to be done?

Right now Australia's climate policy relies on two things working: carbon offsets to soak up the impact of industrial emissions, and for CCS technology to materialise and get us out of the mess we are in. Although they might sound like good ideas in theory, the problem is that in practice these things don't actually work. The Australian government has a long history of relying heavily on the land sector to demonstrate progress towards reducing emissions while continuing to export vast amounts of coal and gas to the rest of the world. This approach was engineered by John Howard's Coalition government to allow Australia to meet its international obligations under the Kyoto Protocol, which ran from 2005 until 2020, when it was superseded by the Paris Agreement. Creative accounting has been used by Australia to calculate the natural uptake of carbon by vegetation and soils to obscure our government's continuous support for the fossil fuel industry. Under the ALP's signature climate policy, the Safeguard Mechanism allows Australia's largest industrial polluters to buy "carbon credits" to balance out their emissions, instead of actually requiring them to reduce the huge volume of carbon they are dumping into the atmosphere for free. The argument can seem a bit confusing at times – that's because it's designed to fly under the radar. So while the details are a bit mind-numbing, please bear with me. Given that carbon offsets are currently being used to prop up Australia's climate policy in the short term, and CCS is the world's long-term plan to reduce emissions, it's important to have a basic grasp of what's going on. Because if we buy into these delusions, we will be in very deep trouble.

First, let's look at the issues with carbon offsets. When you carefully examine the latest greenhouse gas figures, they show that nearly all of the reductions in Australia's emissions are currently coming from electricity and land use changes. The reported 5 per cent decline in electricity emissions is due to the uptake of renewable energy displacing fossil fuel power sources. Although this is good progress worth celebrating, unfortunately right now the land sector is doing all the heavy lifting. Reductions in land-clearing, including avoiding the logging of native forests, are being used to generate carbon credits through the Australian Carbon Credit Unit (ACCU) scheme that then "offset" the emissions of big polluters through the Safeguard Mechanism. The details are convoluted, but basically heavy industries can effectively carry on with business as usual as long as they pay to plant a few trees or avoid logging to cancel out their industrial emissions. In this way the land-use sector can be counted as a "net sink" of carbon emissions, which effectively brings emissions down when calculating our national greenhouse gas figures. The latest numbers show that the land sector removed more carbon dioxide from the atmosphere than it released, contributing to the drawdown (or negative emissions) of around 64 million tonnes of carbon dioxide equivalent in 2023.

Australia's use of the land sector has been criticised internationally for creating an illusion of progress while industrial emissions continue to rise. Independent analysis by Climate Action Tracker explains how our failure to genuinely reduce our emissions is masked by the Australian government's revision of its carbon accounting method. In essence, the government uses a reference year of 2005, a year when Australia had unusually high levels of land-clearing. If you use a baseline when you cleared a record amount of land, then it looks like emissions are falling because of the high starting point. As policy analyst Polly Hemming explains: "It's a bit like saying because you don't drink on average as much alcohol as you did on the night of your 21st birthday, that means you have reduced your alcohol consumption."

When these adjustments in land use are incorporated into 2030 projections, it's easier for Australia to meet its climate targets. As Climate Action Tracker points out, based on these revised calculations, other emissions only need to drop by 24 per cent. Before the revision, the decline was 32 per cent. This dodgy accounting serves to hide the need to reduce emissions across all polluting sectors, and it works. According to the Australia Institute, if you remove the inflated land-use figures and just count actual emissions from polluting industries and the like, Australia's emissions have only declined by around 1.2 per cent since 2005. Obviously this falls well short of meeting the government's legislated target of a 43 per cent reduction in emissions by 2030. But if you include land use, the government claims that Australia's emissions have fallen 24.5 per cent since 2005, masking the negligible reductions in total emissions across all sectors.

The Climate Change Authority has reported that Australia's emissions are still only declining at around half the required rate to achieve our 2030 target, with our efforts to reduce total emissions currently fluctuating around zero. The government's plans for achieving our international obligations under the Paris Agreement rely heavily on meeting the 82 per cent renewable electricity goal. This means increasing the share of renewable power from 36 per cent to 82 per cent in less than six years. In other sectors of the economy, emissions have mostly been flat, or, in the case of transport and agriculture, are actually rising. Independent analysis of the likelihood of achieving net zero by 2050 in the absence of radical policy changes suggests that, on our current trajectory, Australia is unlikely to get much below a 50 per cent reduction on 2005 levels by 2050, let alone achieve net zero.

To make matters worse, there are serious concerns about the integrity of the Australian carbon credits scheme being used by heavy polluters like gas plants and coalmines to meet their obligations under the Safeguard Mechanism. Legal professor Andrew Macintosh, the former chair of the Australian Government's Domestic Offsets Integrity Committee and Emissions Reduction Assurance Committee, has argued that many of the carbon credits being used to offset industrial emissions have very low integrity. In an explainer for *The Conversation* he writes: "People are getting carbon credits for not clearing forests that were never going to be cleared anyway, for growing trees that already exist, for growing forests in places that will never sustain them, and for operating electricity generators at landfills that would have operated anyway." In other words, Australia's carbon offsets system is a sham. Using a flawed scheme that heavily relies on the land sector, which cannot be considered a permanent solution in a continent as bushfire- and drought-prone as Australia, simply delays doing the real work of genuinely reducing emissions. We can run, but ultimately we can't hide from the fact that dubious carbon offsets are not a serious solution to reducing emissions. The longer we delude ourselves, the more warming we lock in as emissions from fossil fuels continue to accumulate in the atmosphere and further destabilise the planet.

*

Since carbon capture and storage (CCS) is being proposed as a silver bullet for reducing emissions around the world, it's worth doing a quick reality check on the state of the industry. The UNEP reports that around 80 per cent of pilot CCS projects over the past thirty years have failed, with operational projects storing less than 10 million tonnes of carbon dioxide per year. But according to the Global CCS Institute industry group, in 2023 there were forty-one commercial CCS plants worldwide capable of capturing 49 million tonnes of carbon dioxide each year. Note that this is five times the estimate independently reported by the UNEP, suggesting the figures are being overstated by the industry. But for argument's sake, let's take the industry's higher estimate and assess the technology's ability to solve our climate woes. If we compare the 49 million tonnes sequestered by CCS to the record high 40.6 billion tonnes of carbon dioxide the world burnt through in 2023, currently operational carbon capture plants are only able to offset approximately 0.1 per cent of global carbon emissions. In other words, CCS technology is currently only offsetting one-tenth of 1 per cent of total global carbon emissions emitted each year.

For CCS to be a viable solution, operational capacity would need to be scaled up roughly 1000 times to offset current global emissions. By the industry's own admission: "We are now seeing a rapid escalation in the development of new CCS projects, although relatively few have yet advanced to operation." To reach net zero by 2050, the industry group estimates that between US$655 billion and US$1.3 trillion needs to be invested in CCS technology to make this a reality. Aside from the trillion-dollar price tag, it's important to realise that CCS projects take around ten years to progress through concept, feasibility, design and construction phases before becoming operational – time we simply don't have.

It is important to understand that billions of tonnes of carbon dioxide already in the atmosphere will need to be removed to achieve the Paris Agreement targets. But there is a critical distinction to flag here – carbon removal technology is needed to try to draw down *existing* emissions from the atmosphere, rather than being misused as a justification for the generation of *new* emissions.

As things stand, the world needs to reduce greenhouse gas emissions 43 per cent by 2030 and 60 per cent by 2035 relative to 2019 levels to limit warming to 1.5°C with no or limited overshoot of that target. And yet the global stocktake conducted at COP28 concluded that if all nationally determined contributions, including all conditional elements that require funding, technology and capacity building, are *fully implemented*, then global greenhouse gas emissions are projected to decrease just 5.3 per cent from 2019 levels by 2030. So, in other words under the best-case scenario, collectively we are on track to reduce emissions by around 10 per cent of what the science says is needed by 2030. The global stocktake text also "notes with concern" that the implementation of *current pledges* would only reduce emissions by around 2 per cent instead of the 43 per cent needed by 2030. This means that, right now, our collective action will, at best, only shave off less than 5 per cent of the reduction in global emissions required by 2030.

So despite all the political hot air, we are disastrously off-track to decarbonise our world, because CCS is being used to delay global progress in shutting down the fossil fuel industry. Given the potential failure of CCS technology to become sufficiently viable, the United Nations Environmental Programme calls for countries to aim for "a near total phase-out of coal production and use by 2040 and a combined reduction in oil and gas production and use by three-quarters by 2050 from 2020 levels, at a minimum." The amount of carbon dioxide removal needed in the second half of the century will only be feasible if we see substantial new deployment of CCS technology within the next ten years. To achieve global targets, approximately 1 billion tonnes of carbon dioxide need to be stored each year by 2030, growing to 10 billion tonnes per annum by 2050. But do we want to divert limited financial and environmental resources into propping up the fossil fuel industry until the bitter end, or are we finally ready to get serious and invest in renewable energy technology that already exists and could be rolled out in a handful of years? Any sensible person can see that pinning our hopes on expensive, unproven technology that will take decades to work at the scale required is absolute madness.

*

To its credit, the Albanese government has tried to support Australia's emergence as a renewable energy superpower. Since the ALP's election win in May 2022, we have seen a Net Zero Economy Agency created to manage the decarbonisation of the economy; expansion of the Capacity Investment Scheme to support the roll-out of dispatchable renewable energy and battery storage; and a National Reconstruction Fund to invest in renewable infrastructure and low-emissions technologies to boost Australia's future industrial capabilities. While these are all steps in the right direction, the challenge is not to undo all of this good work by allowing the interests of the fossil fuel industry to co-opt the process and weaken real progress towards reducing global emissions.

Rather than "net zero," the goal must be to achieve "real zero," which can only happen once we stop burning fossil fuels. In fact, the science tells us that around 60 per cent of oil and gas reserves and 90 per cent of coal must remain unextracted if warming is to be limited to 1.5°C. There is no way around having to eventually face this scientific reality.

But instead of facing facts, in December 2023, the federal government caved in to lobbying from the oil and gas company Santos. A series of emails between Santos and the Department of Climate Change, Energy, the Environment and Water released through the freedom-of-information process revealed relentless pressure from the fossil fuel giant to make amendments to Australia's sea-dumping laws, known as the *Environment Protection (Sea Dumping)*

Act 1981. Industry lobbyists called for changes to allow carbon dioxide to be pumped into international waters to offset emissions from new fossil fuel projects such as Santos's Barossa offshore gasfield, which plans to drill for gas in the Timor Sea and pipe it more than 200 kilometres to Darwin for processing. The most significant amendments related to Article 6 of the London Protocol, which previously prohibited "the export of wastes or other matter to other countries for dumping or incineration at sea" and similarly prohibited the transboundary transportation of carbon dioxide for geological storage. These changes removed the final barriers for the industry to use CCS as a way of justifying the expansion of fossil fuel production in our region.

With the passage of these amendments through the Senate in November 2023, Australia's environmental minister can now issue permits for giant fossil fuel projects to export carbon waste from their operations and bury millions of tonnes of carbon under the sea floor beyond Australia's national boundaries. While Santos has billed the Bayu Undan depleted gasfield as a CCS storage hub for the Asia-Pacific region, the company's own plans involve using CCS to reduce emissions from its controversial Barossa offshore gasfield development, which contains nearly double the carbon emissions of any other gas project in Australia. The amendments also pave the way for countries such as Japan and Korea to export carbon dioxide to Australia to be dumped offshore. The fossil fuel industry's vision for our nation is disturbing: if we can't continue to be a quarry, then we will be a dumping ground. It's a move that also doesn't bode well for poorer, developing nations: with less stringent environmental regulations, they are vulnerable to exploitation by wealthier nations refusing to reduce their use and production of fossil fuels.

Relying on unproven CCS technology to reduce emissions, even on land, is a very risky prospect. Offshore CCS has added dangers of acidifying marine environments, contaminating groundwater, inducing earthquakes and the displacement of toxic brine deposits. The true risks of the hazards of the offshore CCS industry are yet to be fully scientifically and technically assessed, let alone comprehensively regulated. In a 2023 report, the

Center for International Environmental Law warns: "The push for offshore CCS reflects the same attitude that has left the oceans in crisis today: treating them as a limitless resource to exploit and a bottomless receptacle for humanity's waste." If plans to construct more than fifty new offshore CCS projects around the world go ahead, it would represent a 200-fold increase in the amount of carbon dioxide currently being injected under the seabed each year. While this might sound like a major increase, collectively these projects would theoretically capture only around 450 million tonnes of carbon, or just 1.5 per cent of current annual global emissions from energy and industry. Given that the history of the offshore oil and gas activity is full of examples of deadly accidents and environmental disasters associated with spills and leaks, embarking on such a risky path for such little gain is spectacularly illogical.

Most of the proposed offshore CCS projects plan to capture emissions, then transport them by pipeline to an offshore storage site, mostly using depleted oil and gas wells. Underwater pipelines and oil shipping leaks are already so common that one recent study of satellite images between 2014 and 2019 found enough oil slicks floating on the ocean surface to cover an area close to the size of Queensland. If the oil and gas industry is to continue expanding, we can expect this problem to get much worse. Once decommissioned, pipelines and other offshore equipment are often left unmonitored at the bottom of the sea. So after decades of plundering the seas for oil and gas, fossil fuel companies are now planning to use the ocean floor as a final dumping ground for their waste on an unprecedented industrial scale. Many of the proposed offshore CCS projects are planning to collect carbon emissions from multiple industrial facilities and inject them into shared subsea storage hubs – an approach that has never been tested. Until now, global experience with offshore CCS has been based on just two projects in Norway, both of which encountered unpredicted problems despite their relative simplicity and small scale. Injecting massive amounts of carbon dioxide under the seabed presents an uncalculated risk that has never been confronted at the scale being proposed by the fossil fuel industry.

Aside from the risk of further acidifying the oceans, which causes great harm to marine ecosystems, there is the added complexity of trying to manage reservoir pressure and monitor risks in the deep ocean. And if the environmental risks aren't enough, the multi-billion-dollar price tag should be enough to deter our leaders from going down this time-wasting track. As the Center for International Environmental Law explains, without a market for storing carbon or a meaningful penalty for failing to do so, CCS is just an expense that the industry is counting on governments to subsidise. Already, the federal government is providing $141.1 million over ten years to support CCS projects, including infrastructure to support the expansion of the offshore oil and gas industry, particularly in Western Australia and the Northern Territory.

When you take a step back and realistically assess the integrity of Australia's climate policy, it's clear that our government is still continuing to support the expansion of the fossil fuel industry by approving new carbon-intensive developments, heavily subsidising polluting industries and pinning its hopes on non-existent CCS technology and junk carbon credits to save us. If we fail to meet our targets because we buy into a delusion that carbon offsets and CCS will reduce Australia's emissions, this disastrous delay will make the job so much harder in the future and bake in higher levels of global warming for centuries to come. If we are serious about becoming the clean energy superpower we know Australia can be, our political leaders need to be brave enough to do the right thing and stop approving and subsidising any new coal, oil and gas projects. As we are seeing right now, gains made by increasing renewable power are still being swamped by fossil fuel emissions and our inability to soak up carbon fast enough. We have to be honest with ourselves and admit that what we are doing isn't working. Are we willing to accept another wasted decade of greenwashing as the planet burns?

When dealing with a medical emergency, it's a case of first things first. We have to stabilise the situation and minimise the influence of any factors that might do more harm. The same applies to addressing climate change. Given that the primary cause of our warming planet is the burning of fossil fuels, the first thing we need to do to limit global warming is immediately stop subsidising the production and use of coal, oil and gas. According to the International Monetary Fund (IMF), fossil fuels in most countries are priced incorrectly as they do not reflect the full social costs associated with supply, environmental degradation, carbon emissions, local air pollution and general taxes applied to consumer goods. And in 2022, direct fossil fuel subsidies more than doubled globally, from half a trillion US dollars to US$1.3 trillion. When you factor in indirect subsidies that include the costs imposed by undercharging for environmental and social impacts like premature deaths from exposure to air pollution and lost government revenue from under-taxing polluters, it brings the total subsidy figure to US$7 trillion in 2022, or around 7 per cent of global economic output as measured by gross domestic product.

The Australia Institute has calculated that in the 2022/23 financial year, Australian federal and state governments provided $11.1 billion in subsidies to the fossil fuel industry. The fuel tax credit scheme, which refunds tax paid on the use of diesel and petrol to support large mining projects, was the big-ticket item, accounting for $8 billion. That's more than government spending on the Australian Army, which, ironically, is increasingly being called on to respond to natural disasters. On that point, fossil fuel subsidies are projected to reach a record $57.1 billion over the lifetime of projects planned by the government, which is nearly fifteen times greater than the $3.9 billion available in the Disaster Ready Fund, the nation's emergency-response budget.

To pour even more fuel on an already raging fire, the Albanese government continues to subsidise gas infrastructure, with $1.9 billion going towards the Middle Arm petrochemical hub in Darwin. A further $141.1 million over ten years will be directed to CCS projects and $217 million to build roads specifically for the gas industry in the Northern Territory. It's a stunning amount to be diverting into infrastructure that will expand the operation of the fossil fuel industry during a climate emergency. The Middle Arm precinct will provide major export opportunities for unconventional "fracked" gas from the Beetaloo Basin, a project with the potential to contribute up to 1.2 billion tonnes of carbon until 2050. In comparison, Australia emitted 460 million tonnes of greenhouse gases in the year ending June 2023, so this single operation could contribute 2.5 times Australia's total annual emissions over the lifetime of the project. Research by Climate Analytics suggests that domestic emissions from fracking in the Beetaloo and processing at Darwin's Middle Arm industrial precinct would produce up to 49 million tonnes of carbon dioxide equivalent per year, around 10 per cent of Australia's total emissions. That means a single project would produce more emissions than the entire reduction goal under Labor's Safeguard Mechanism. Instead of making real and sustained cuts to emissions, Australia is supporting fossil fuel projects that will release carbon that will still be in the atmosphere thousands of years from now.

Underpricing the true cost of fossil fuels means that governments forgo a valuable source of revenue needed to fund the clean energy transition, climate change adaptation and the repair of carbon-absorbing ecosystems. Instead, the profits continue to fatten the coffers of private corporations. For example, the world's five largest oil companies – BP, Shell, Chevron, ExxonMobil and TotalEnergies – have made more than a quarter of a trillion dollars since Russia's invasion of the Ukraine in February 2022, when the supply of Russian gas and oil was withheld from global markets. Because Russia is a major exporter of fossil fuels, accounting for close to a quarter of the world's gas and 12 per cent of oil, disruption to the global supply chain delivered a massive shock to energy prices, driving up inflation and placing cost-of-living pressures associated with heating, cooling and lighting on vulnerable households. The ongoing energy crisis has highlighted the risk of overdependency on fossil fuels – especially those sourced from hostile nations – at a time when the world should be rapidly transitioning to clean energy. It also makes a strong case for the decentralisation of electricity generation in helping address societal inequalities.

For renewable power to be scaled up at the rate needed to displace global reliance on coal, oil and gas, incentives are needed to support the clean energy transition while penalising polluting industries for the damage they cause. While the case for pricing carbon makes perfect sense, right now the exact opposite is the case. Undercharging for oil products accounts for nearly half of global fossil fuel subsidies, coal a further 30 per cent, and gas another 20 per cent. The IMF argues that a full reform of fossil fuel prices that removes direct fuel subsidies and imposes a corrective carbon tax that makes polluters pay for the environmental and social damage caused by their industry could reduce carbon emissions 34 per cent below 2019 levels by 2030. Remember that to stabilise global warming to 1.5°C with limited or no overshoot, we need to reduce emissions by 43 per cent by 2030, so in a perfect world, removing global fossil fuel subsidies would get us most of the way there.

If our political leaders truly wanted to address the worsening climate crisis, these types of economic measures could be done very quickly. It would also avoid wasting precious time waiting for expensive technological solutions that may take a decade or more to materialise, if they eventuate at all. The IMF explains that a full transformation of fossil fuel pricing could raise revenue worth US$4.4 trillion in 2030, or around 3.6 per cent of global GDP. Such measures would create the global economic conditions that would genuinely incentivise the rapid phase-out of the polluting industries that are destroying the planet.

One of the key findings of the IPCC's Sixth *Assessment Report* that seems to have gone unnoticed is the fact that there are solutions available *right now* across all sectors of the economy that could at least *halve* greenhouse gas emissions by 2030. The International Energy Agency's *Net Zero Roadmap* is even more optimistic, suggesting that ramping up renewables, improving energy efficiency, reducing methane emissions and increasing electrification with technologies available today could deliver more than 80 per cent of the emissions reductions needed by 2030. The biggest wins are in the transition to solar and wind power, where markets have seen an 85 per cent drop in the price of solar cells, wind turbines and battery technology since 2010. Together, solar and wind power are the cheapest way of reducing emissions, costing less than US$20 per tonne of carbon dioxide. That's at least five times cheaper than the IPCC's estimate of US$100 to US$200 per tonne needed for CCS, with renewables having the added advantage of being proven technology that can be deployed very quickly. There's no need to wait another decade before we get the drop in emissions that we need to stabilise the Earth's climate. It's a no-brainer that would deliver environmental, social and economic benefits within the very short timeline we have to limit warming to below 2°C.

In the words of the chair of the IPCC, Hoesung Lee:

> We are at a crossroads. The decisions we make now can secure a liveable future. We have the tools and know-how required to limit warming. I am encouraged by climate action being taken in many countries. There are policies, regulations and market instruments that are proving effective. If these are scaled up and applied more widely and equitably, they can support deep emissions reductions and stimulate innovation.

The message from the scientific community couldn't be clearer: achieving the goals of the Paris Agreement is technically possible, but only if the political will is there. One of the few encouraging outcomes of the COP28 summit was an agreement to triple the world's renewable energy capacity by 2030. Not only will this new capacity displace the need for coal, oil and gas, it will also support areas such as transport that can be "electrified" by switching from petrol to electric vehicles. The International Energy Agency has shown that the tripling of renewables would halve the need for coal power and deliver the added benefit of halving the "fugitive" methane emissions generated from coalmining. Stringent and effective policies that favour the rapid deployment of renewable power would slash fossil fuel demand 25 per cent by 2030 and a huge 80 per cent by 2050. Rapid increases in clean energy capacity remove the need for any investment in new coal, oil and gas projects, and would also provide enough time for an orderly phase-out of existing fossil fuel production.

Solar is expected to be the key technology needed to achieve the target of tripling renewable power worldwide, accounting for two-thirds of the rise in clean energy out to 2030. To achieve this goal, solar capacity across the world needs to increase five-fold this decade, while wind power needs to be scaled up three-fold. Together, solar and wind would bring us over 90 per cent of the way to achieving the tripling of renewables by 2030. Currently only around 30 per cent of the world's electricity is powered by renewables, suggesting that even though it is the cheapest source of energy and the quickest way to reduce emissions, there are still political barriers that need to be overcome. The sheer scale of global fossil fuel subsidies has hampered the speed of the clean energy transition, something that will be considered morally reprehensible in the future when people look back and see the situation for what it was – polluting industries blocking the rise of the clean industrial revolution to protect corporate profits, no matter the social and environmental cost.

Although the world invested a record US$1.8 trillion in clean energy in 2023, this needs to climb to around $4.5 trillion each year by the early 2030s to keep the Paris Agreement targets within reach. That might sound steep, but if we want a habitable planet, diverting even half of the $7 trillion spent globally on fossil fuel subsidies would be the obvious way forward. The money is there; we just need to get our priorities straight. When agreeing to spend more than a quarter of a trillion dollars on nuclear-powered submarines, former Coalition treasurer Josh Frydenberg said: "Everything is affordable if it's a priority." Clearly, addressing climate change is still not a priority.

Despite the strength of the scientific case for the switch to renewables, and Australia's natural advantages of abundant sunshine due to our subtropical location, and our ability to harness the power of the westerly wind belt, our potential as a renewable energy superpower is still largely untapped. Even though we are the sunniest continent on the planet, less than 15 per cent of our electricity is generated by solar power. The good news is that, aside from Western Australia, all of Australia's states and territories now have emissions reduction and renewable energy targets of some kind. Perhaps our biggest clean energy success story is South Australia, which is on track to reach 100 per cent renewable electricity generation as soon as 2027, bringing forward its renewable energy target by three years. In just over twenty years, the state has gone from 1 per cent of its electricity generated from solar and wind power to around 70 per cent renewables today.

Currently South Australia generates around a quarter of its electricity from solar and 44 per cent from wind farms, with five grid-scale batteries in operation and two more under construction, with plans to be an exporter of excess energy to the more densely populated eastern states. In the summer of 2022–23, a record 80 per cent of the state's electricity was generated by solar and wind, including a ten-day period when the state ran on 100 per cent green energy. Such high capacity opens up the possibility of supporting the development of the energy-intensive hydrogen industry that will be needed to decarbonise hard to abate sectors such as steel, aluminium, ammonia, chemical production and heavy transport, which have traditionally been powered by fossil fuels. To support its emerging green manufacturing industry, in 2023 the South Australian government introduced the Hydrogen and Renewable Energy Bill to support the development of the hydrogen industry and renewable energy generation, highlighting how important political vision and leadership is at this critical crossroads.

The prospect of producing "green steel" using hydrogen-powered renewable energy represents a huge economic opportunity for Australia to be a leader in the green industrial revolution currently underway. In a 2023 study that assessed the potential of different parts of the world to support a globally competitive green hydrogen industry using high-quality renewable energy and iron ore deposits, Australia was highlighted as a potential global leader of green hydrogen-based steel manufacturing. Western Australia was singled out as the area of highest potential for green steel manufacturing because of the availability of iron ore reserves and a stable investment environment, compared with countries such as Ukraine, Iran and Peru. Instead of being the world's largest exporter of raw iron ore to countries like China, Australia could instead become one of the biggest exporters of green steel, creating new manufacturing opportunities that will not only reduce emissions domestically and internationally, but would also create jobs needed for a zero-emissions economy. If we can put aside the hysteria and non-scientific arguments against renewables, we will see that a world of opportunity awaits us.

*

Instead of the "polluter pays" principle, right now Australian taxpayers are footing the bill to keep the fossil fuel industry on life support. According to the Australia Institute, the government collects more money from the Higher Education Contribution Scheme (HECS) than it does from the Petroleum Resource Rent Tax (PRRT), a federal tax levied on petroleum, oil and gas projects in Australian waters. In 2022/23 the government collected nearly $2.3 billion from the PRRT, less than half the $4.9 billion generated from student loan repayments. In his National Press Club address in January 2024, Richard Denniss, executive director of the Australia Institute, said: "Consider the fact that in Norway, they tax the fossil fuel industry and give kids free university education, in Australia we subsidise the fossil fuel industry and charge kids a fortune to go to university." And it's not just young people doing it tough. In 2020/21, nurses paid more than three times the income tax than the gas industry paid in income tax and the PRRT combined. Clearly we have things backwards.

Despite the fact that eminent economists like Ross Garnaut have been arguing for over a decade that placing a price on carbon is the most efficient way of driving down emissions, Australian political history shows that entrenched fossil fuel lobbyists will fight tooth and nail to preserve the status quo. Most recently, in February 2024, Garnaut and fellow economist

Rod Sims proposed introducing a carbon price on all fossil fuel extraction sites and fossil fuel imports to Australia. Their "carbon solutions levy" could generate over $100 billion in its first year, providing vital funds to accelerate of Australia's renewable energy roll-out and subsidise the development of low-carbon manufacturing for products like green steel and aluminium. It's a blindingly obvious way forward, and yet the plan was immediately rejected by the federal government and the Nationals, but endorsed by the Greens. Despite the escalating climate crisis, it's clear that we still have a long way to go to end the climate wars in Australia. Irrational ideology is still undermining our democracy and wasting the last of the precious time we have left to secure a liveable future.

Continuing to support false solutions that prop up the fossil fuel industry simply diverts limited resources away from proven technologies that can reduce emissions quickly and cheaply, with far less risk to communities and the environment. The CSIRO's *Pathways to Net Zero Emissions* report shows that Australia could halve its emissions by 2030 using existing technologies. It identifies decarbonisation of the electricity sector as the way to unlock Australia's clean energy transition in a range of other sectors like residential and commercial buildings, transport and heavy industry. By 2030, a tripling of renewables would see solar and wind account for three-quarters of electricity generation. In the CSIRO's modelling, fossil fuel use in the electricity sector would fall from today's figure of 70 per cent to less than 10 per cent in 2030, and would be almost entirely eliminated by 2040 except for gas "peaking plants" that only get switched on 1 to 5 per cent of the time to ensure reliable supply during times of high demand and low solar and wind production.

While there may be a role for gas peaking plants as we wean off fossil fuels and ramp up the roll-out of renewables, analysis by the Clean Energy Council shows that ultimately large-scale battery storage is a superior, more flexible alternative to gas turbines for meeting peak electricity demand. Battery storage outcompetes emissions-intensive gas on cost, faster reaction time and easier deployment, with the obvious benefit of zero emissions.

Forget doubling down on gas and CCS, battery storage is the true bridge to Australia's clean energy future.

The CSIRO has also demonstrated that electricity produced by renewables is cheaper than fossil fuel power generation and is expected to remain the lowest-cost power source for decades to come. Even with storage and transmission factored in, renewable electricity generation from solar and wind today costs between $112 per megawatt hour, the unit used to measure electric output. In contrast, the cost of nuclear power generated using small modular reactors – the approach currently being spruiked by the Coalition – is $509 per megawatt hour. Investing in renewables is clearly a complete no-brainer. The CSIRO has concluded that large-scale nuclear reactors are not appropriate for Australia's relatively small electricity grids, and the time needed for small modular reactors to prove commercially viable "rules it out of any major role in the electricity sector emission abatement required for Australia to reach its net-zero emissions target in 2050."

The Coalition's attempt to resurrect the nuclear power debate in Australia is simply a red herring intended to delay the inevitable transition to renewable energy. Despite all the talk and a history of failed experiments, small modular reactors do not exist anywhere in the world and are unlikely to play a part in Australia's energy mix until the 2040s, if at all – even if someone were willing to pay the exorbitant price. Aside from it being the most expensive form of energy, Australia does not have an existing nuclear industry or the legal framework for nuclear power generation. Hypothetically we would have to wait at least ten to twenty years for a nuclear power station to be built, compared to the one to three years it takes to roll out major wind and solar projects. As journalist Adam Morton pointed out in *The Guardian*: "The advice from government agencies and nearly all energy experts is that renewable energy will be able to provide more than 90 per cent of Australia's electricity needs at much cheaper cost long before a nuclear industry could be built from scratch."

And even if we were prepared to pay almost five times more for nuclear power and risk delaying reducing our emissions for another twenty years, the intergenerational environmental and societal risks are completely unacceptable, especially when safer alternatives already exist. You only have to look at the major nuclear disasters at Chernobyl, Three Mile Island and Fukushima, which exposed hundreds of thousands of people to radioactive material and contaminated vast areas of land, to know that nuclear power is not a serious solution to address climate change in a country like ours, which is blessed with the richest solar and wind resources in the world. The Coalition's diversion into a debate about nuclear power, like its promotion of carbon capture and storage, is a cynical effort to undermine the support of renewable energy and extend our reliance on fossil fuels for as long as possible.

Thankfully, more progressive parts of the world are moving on with the inevitable task of decarbonising their economies, so Australia will be dragged into the twenty-first century kicking and screaming if necessary. The CSIRO warns that lagging behind international efforts to decarbonise will put Australia at a competitive disadvantage as other nations increasingly adopt low-emissions technologies and impose trade barriers on high-emitting nations. For example, the European Union has introduced the Carbon Border Adjustment Mechanism, which will impose a tariff on carbon-intensive goods like steel, electricity and hydrogen entering the region to encourage cleaner industrial practices throughout the global supply chain. The executive director of CSIRO's Environment, Energy and Resources division, Peter Mayfield, has said: "The transformation required in our energy system is key, and one of the biggest shifts we will see in our lifetimes. The way forward is not going to be simple or easy. Dragging our feet will almost certainly result in more costly decarbonisation, increased risk of assets becoming stranded or impaired, greater sovereign risk, and the overall long-term competitiveness of our economy being severely hampered."

We know what we need to do and we know how to do it. Our political leaders just need to keep sight of what is sensible and sane during this critical time of transition. We need them to stop chasing unicorns and back the right horse, waiting patiently in the stable.

While I was writing this essay, the fifth mass coral bleaching event to strike the Great Barrier Reef since 2016 began to unfold. The record temperatures of 2023 spilled into 2024, with January, February and March all continuing the global streak of relentless ocean heat that began in June, culminating in the declaration of the fourth planet-wide mass bleaching event on record. Every ocean basin in the world has been affected, with the Great Barrier Reef experiencing the most severe and widespread event in its history. The southern section of the reef, which had previously escaped harm, finally succumbed to sustained heatwave conditions, resulting in extensive bleaching throughout its entire 2300-kilometre range. Early estimates suggest that up to 80 per cent of the Great Barrier Reef has been damaged. From the Torres Strait in the far north to Lord Howe Island, the world's most southerly reef, corals are turning white from heat stress and starting to die. Describing the conditions at Heron Island, off Gladstone in the reef's south, the project manager of Coral Watch at the University of Queensland, Diana Kleine, told *The Guardian*: "It's devastating. Unbelievable. The water was way too warm. Heron has escaped bleaching several times but this year it has hit so hard."

Scientists measure the accumulation of heat stress using a metric called "degree heating weeks," which factors in both the duration and intensity of extreme heat exposure. As eminent coral reef scientist Terry Hughes explains, as a rule of thumb two to four of these heating units can trigger the onset of bleaching, and heat-sensitive species of coral begin to die with conditions between six and eight. Values over ten are considered catastrophic. So far during this event, heat stress on the Great Barrier Reef has climbed to as high as fifteen to eighteen heating units on parts of the reef, well above all previous mass bleaching and mortality events. Experts estimate that half of the Great Barrier Reef's shallow-water corals were killed between 2016 and 2017. It is still unknown – or undisclosed by the government – exactly how much more permanently died off during the 2020 and 2022 mass bleachings, and how this most recent bout of extreme heat will affect our beleaguered reef. As of April 2024, the event is still underway, so the overall mortality figures won't be known for a while, but it is safe to say that the carnage is going to be widespread and unprecedented. We are witnessing the slow death of the largest living organism on the planet. This is how it is going to be if we refuse to care about the future of our country. It is the beginning of the end of life as we've known it.

Writing about the disconnect between climate science and policy while the Great Barrier Reef bleaches is no longer heartbreaking, it is infuriating. Every decision that has ever been made to destroy the natural world has led us to this place. The destruction we are witnessing is the culmination of corporate disregard for the very essence of what makes us human and government refusal to properly regulate polluting industries. It's a moral failure of colossal proportions. We know why global warming is happening and what we need to do to stop it. And yet the world still plans to produce more than double the amount of fossil fuels in 2030 than is consistent with limiting warming to 1.5°C. Here in Australia, there are currently plans for 100 new coal and gas projects. Instead of pausing at this fateful crossroads and choosing to change direction, our leaders are still barrelling down the highway to climate hell.

Since the first IPCC report came out in 1990, the Australian government has spent at least $137 billion subsidising fossil fuel consumption and production that has made the problem of climate change worse. Instead of responding to the repeated warnings of the scientific community, our elected leaders are still making decisions that are doing harm. As a scientist, it has become increasingly hard to know what to do. I've found out the hard way that it is career-limiting to speak out about the scientific reality of climate change. Many of us are worried about government funding and are told to "stay in our lane." Others play it safe, keep their heads down and their mouths shut. Most scientists in the university sector are under immense pressure to take on career-killing teaching and administration workloads that result in countless unpaid hours doing work that used to be done by a team of people in an era when higher education was valued. Researchers in government agencies like the Bureau of Meteorology and the CSIRO have to go through multiple levels of approval before they can communicate with the public, and even then they can only comment in very restricted ways. This diverts experts away from being able to provide timely and frank commentary and advice on vitally important policy areas. Meanwhile, the climate crisis unfolding outside our windows is worsening by the day. But when climate scientists don't speak up, there is always someone far less qualified willing to step in and fill the vacuum, often doing more harm than good.

Those of us who do try to communicate the dangers of unmitigated climate change to the public feel like a broken record. We know exactly how to put the brakes on global warming – the IPCC has been saying it for at least thirty years. We need our politicians to listen and have the heart and the courage to do the right thing. But for them to do what they know is right, ordinary Australians need to care about the future of our planet and vote for leaders who reflect the values of our local communities. As our next federal election approaches, we need a critical mass of people willing to create a social tipping point that demands our leaders do better.

As I wrote in my book *Humanity's Moment*:

> We will not see the political response we need to address climate change until we redefine the cultural and social norms that are destroying life on Earth. Individual voters are responsible for creating or removing the social licence needed to maintain the status quo of burning fossil fuels to the point of planetary instability. As we start to see democracy being undermined around the world, we don't have the luxury of being apolitical. We are at a critical crossroads, where it is important that we don't become disengaged from political processes that will determine our future. We need to do what we can, in our own way, using our own voice, to meet the crisis we face.

Although I've been a research scientist for decades, I'm starting to wonder if any of my work will matter ten or twenty years from now, if we fail to stabilise the Earth's climate. Since serving as an IPCC author, I've been grappling with the dilemma of how best to use my time knowing how bad the situation is – do I keep teaching the next generation of students and chip away at research projects that will incrementally advance the field, or do I try to share what I know as far and wide as possible? And then deep in the winter of 2023 it dawned on me: I've been climbing a ladder that is up against the wrong wall. The most useful thing someone like me can do at this fraught moment in human history is use the time I have left to warn the public. Despite my life-long love of research, I now understand that no amount of extra scientific knowledge is going to bring about the political change we need. Having been through the IPCC process, I am all too aware that our messages are either simply not getting through or are actively being ignored. The extra graphs, datasets and reports just keep piling up as our climate continues to destabilise. We need more climate scientists with enough time to be involved in the political decisions being made about our future right now.

Eventually the dilemma of carrying on with my version of "business as usual" became too hard to ignore. With a heavy heart I resigned from a permanent academic position at one of the most prestigious institutions in the country at the peak of my career. Walking away from what I thought was my dream job was the hardest decision I've ever made, something I never imagined I would do. But given the lack of institutional support and how high the stakes are, I felt I had no choice. In the words of Martin Luther King Jr: "Our lives begin to end the day we become silent about things that matter." As a climate scientist who understands that we only have limited time left to avert disaster, I realised that I can't wait until I have retired to speak out.

When I told a trusted senior colleague what I had done, he shared a pearl of wisdom from Plato: "If you do not take an interest in the affairs of your government, then you are doomed to live under the rule of fools." While it's a little blunt, nothing could be truer of the moment we face. The reality is that most politicians and their advisers do not have a science degree, or even a science education beyond high school. And yet we are entrusting these people with the most important decisions humanity will ever make.

While I don't expect every politician to be a scientist, the very least they could do is listen to what our community has to say and act in the national interest of future generations.

What happens at the next federal election really matters. As the third-largest exporter of fossil fuels, what Australia chooses to do over the next five years is critical for the stabilisation of the Earth's climate. We still have a chance to determine how bad things get and which areas can be saved, but only if we genuinely reduce emissions. The most urgent thing we need to do is phase out our use of fossil fuels and invest heavily in renewables like our life depends on it. Because it does. What we do during the 2020s will make or break humanity's ability to live safely on the planet. We need to listen to our brightest engineers like Saul Griffith, who are at the ready to help "electrify" our nation with renewable power; our visionary business leaders like Andrew Forrest, who can help Australia be a global leader in the clean industrial revolution; our First Nations leaders like Murrawah Johnson, who can teach us how to better care for Country; and economists like Ross Garnaut, on how we can fund the most important economic transition humanity will ever face. We have world-class knowledge and know-how right here in Australia; we can do this if we have the political will.

There are already irreversible changes in the climate system that will be with us for thousands of years. The challenge right now is to minimise future damage. If we go further down the highway to hell and destabilise the ice sheets, it will be impossible to adapt, especially in an arid country like Australia where most of our population lives on the coast. The fate of our world lies in our hands. And when we go to the polls, remember: we can't reverse-engineer our way out of the problem using carbon capture and storage: it's a fool's errand that will only lead to delay and failure. Ditto nuclear power. Our leaders need to have the guts to put a price on carbon and accelerate the clean energy revolution to transform Australia into the renewable energy superpower we know we can be. When the benefits of a green economy start to be realised, we won't look back. Our only regret will be that we didn't start sooner. It's time to make our vision for a safer and more equitable world not just a reality but a legacy that future generations will be proud of.

SOURCES AND ACKNOWLEDGEMENTS

The author was supported by The Australia Institute's Writer in Residence program and the inaugural Varuna Climate Fellowship. This work was written on Bundjalung country in northern New South Wales, and the traditional lands of the Ngunnawal and Ngambri clans of Canberra, and the Dharug and Gundungurra peoples of the Blue Mountains.

1   "human influence on the climate": Intergovernmental Panel on Climate Change, "Technical summary", *Climate Change 2021: The Physical Science Basis. Contribution of Working Group I to the Sixth Assessment Report of the Intergovernmental Panel on Climate Change*, IPCC, accessed at ipcc.ch/report/ar6/wg1.

2–3   "When experts fail": Joëlle Gergis, "The summer ahead", *The Monthly*, September 2023.

5   "our largest exporters": Australian Labor Party, *Powering Australia: Labor's plan to create jobs, cut power bills and reduce emissions by boosting renewable energy*, 2021, accessed at assets.nationbuilder.com/lean/pages/530/attachments/original/1644715432/PoweringAustralia_ALP_policy.pdf?1644715432.

7   82 per cent of carbon dioxide emissions since 1960: P. Friedlingstein et al., "Global Carbon Budget 2023", *Earth System Science Data*, vol. 15, no. 12, 2023.

8   only 15 per cent think it is an "extremely serious" problem: S. Deshpande et al., *Griffith Climate Action Survey, 2022: Summary for policy and decision making*, Griffith University, 2023, accessed at griffith.edu.au/research/climate-action/national-longitudinal-survey.

8   this polling also showed a disturbing lack of awareness: Institut Public de Sondage d'Opinion Secteur, "Global views on climate change November 2023", 2023, accessed at ipsos.com/en/seven-in-ten-people-anticipate-climate-change-will-have-severe-effect-their-area-within-next-ten-years.

10   "The cumulative scientific evidence": Intergovernmental Panel on Climate Change, *Climate Change 2022: Impacts, Adaptation and Vulnerability. Contribution of Working Group II to the Sixth Assessment Report of the Intergovernmental Panel on Climate Change*, 2022, accessed at report.ipcc.ch/ar6/wg2/IPCC_AR6_WGII_FullReport.pdf.

11   "huge margin": World Meteorological Organization, "WMO confirms that 2023 smashes global temperature record", 2024, accessed at wmo.int/news/media-centre/wmo-confirms-2023-smashes-global-temperature-record.

13   In 2023 the heat stored in the ocean: Carbon Brief, "State of the Climate: 2023 smashes records for surface temperature and ocean heat", 2024, accessed at carbonbrief.org/state-of-the-climate-2023-smashes-records-for-surface-temperature-and-ocean-heat.

12   "2023 was basically": Tyne Logan, "Megafires are increasing with climate change, experts say – but could the emissions they pump out change the climate?", ABC News, 22 January 2024.

13   from April to December the global ocean: World Meteorological Organization, *State of the Global Climate 2023*, 2024, accessed at library.wmo.int/records/item/68835-state-of-the-global-climate-2023.

14   "I was not prepared": Graham Readfearn, "'Huge' coral bleaching unfolding across the Americas prompts fears of global tragedy", *The Guardian*, 11 August 2023.

15   "If global leaders": International Cryosphere Climate Initiative, *State of the Cryosphere 2023: Two degrees is too high*, 2023, accessed at iccinet.org/statecryo23.

16   "this insanity cannot": International Cryosphere Climate Initiative, *State of the Cryosphere 2023*.

16–17 "It's humbling", "the world will be": Gavin Schmidt, "Climate models can't explain 2023's huge heat anomaly – we could be in uncharted territory", *Nature*, vol. 627, no. 467, 2024.

18   the chance of limiting global warming to 1.5°C is just 14 per cent: United Nations Environment Programme, *The Emissions Gap Report 2023*, Nairobi, Kenya, 2024, accessed at unep.org/resources/emissions-gap-report-2023.

23   "under a scenario": M.M. Boer, V. Resco de Dios and R.A. Bradstock, "Unprecedented burn area of Australian mega forest fires", *Nature Climate Change*, vol. 10, 2020, pp. 171–2.

27–8 more than $226 billion: Department of Climate Change and Energy Efficiency, *Climate Change Risks to Coastal Buildings and Infrastructure*, accessed at dcceew.gov.au/sites/default/files/documents/risks-coastal-buildings.pdf.

28   a quarter of Melbourne's Port Phillip council area: Department of Energy, Environment and Climate Action, *Port Phillip Bay Coastal Hazard Assessment*, 2024, accessed at marineandcoasts.vic.gov.au/coastal-programs/port-phillip-bay-coastal-hazard-assessment.

28   online mapping tool: Coastal Risk Australia, "Predicted coastal flooding resulting from climate change NGIS", 2024, accessed at coastalrisk.com.au/home.

29   "The effects of tipping points": CSIRO, "Understanding the risks to Australia from global climate tipping points", 2024, accessed at csiro.au/en/news/All/Articles/2024/February/climate-tipping-points-Australia.

30   "Under high levels of warming": Australian Academy of Science, "The risks to Australia of a 3°C warmer world", 2021, accessed at science.org.au/warmerworld.

34   over half a million properties: Climate Council, "Uninsurable nation: Australia's most climate-vulnerable places", accessed at climatecouncil.org.au/resources/uninsurable-nation-australias-most-climate-vulnerable-places.

35   at least 410 million people: A. Hooijer and R. Vernimmen, "Global LiDAR land elevation data reveal greatest sea-level rise vulnerability in the tropics", *Nature Communications*, vol. 12, no. 1, 2021.

35   approximately 1 billion people: Intergovernmental Panel on Climate Change, "Summary for Policymakers" in *Climate Change 2022: Impacts, Adaptation and Vulnerability. Contribution of Working Group II to the Sixth Assessment Report of the Intergovernmental Panel on Climate Change*, 2022, accessed at ipcc.ch/report/ar6/wg2.

35   up to 560,000 people: Intergovernmental Panel on Climate Change, "Chapter 15: Small Islands" in *Climate Change 2022: Impacts, Adaptation and Vulnerability. Contribution of Working Group II to the Sixth Assessment Report of the Intergovernmental Panel on Climate Change*, accessed at ipcc.ch/report/ar6/wg2/chapter/chapter-15/

35   "Oil and gas projects remain": Department of Industry, Science and Resources, "Resources and energy major projects 2023", Australian Government, Canberra, accessed at industry.gov.au/publications/resources-and-energy-major-projects-2023.

36   "Words like these": Peter Kalmus, "COP out: Wrapping up a useless climate summit that should fool nobody", *Newsweek*, 18 December 2023.

36   "heartfelt congratulations": Organization of the Petroleum Exporting Countries, 4th High-level Meeting of the GECF-OPEC Energy Dialogue, OPEC, 13 December 2023.

37   "We should abandon": O. Milman, "World's top fossil-fuel bosses deride efforts to move away from oil and gas", *The Guardian*, 21 March 2024.

37   world governments are planning: Stockholm Environment Institute, *Production Gap Report 2023: Phasing down or phasing up? Top fossil fuel producers plan even more extraction despite climate promises*, United Nations Environment Programme, 2023, accessed at unep.org/resources/production-gap-report-2023.

37   just five countries: Oil Change International, *Planet Wreckers: How countries' oil and gas extraction plans risk locking in climate chaos*, 2023, accessed at priceofoil.org/2023/09/12/planet-wreckers-how-20-countries-oil-and-gas-extraction-plans-risk-locking-in-climate-chaos.

38   "business as usual": Madeleine King, *Resources Statement to Parliament*, Australian Government, 29 March 2023, accessed at minister.industry.gov.au/ministers/king/speeches/resources-statement-parliament.

38   power the equivalent of the United States and Canada: International Energy Agency, *Renewables 2023: Analysis and forecasts to 2028*, 2023, accessed at iea.org/reports/renewables-2023.

39    $6.7 billion in 2022: Clean Energy Council, *Clean Energy Australia Report 2023*, accessed at cleanenergycouncil.org.au/resources/resources-hub/clean-energy-australia-report.

39   around 90 per cent of coal and three-quarters of all gas: Department of Climate Change, Energy, the Environment and Water, *Australian Energy Update* 2023, accessed at energy.gov.au/publications/australian-energy-update-2023.

39   "Customers of": Climate Change Authority, "2035 emissions reduction targets", 2024, accessed at climatechangeauthority.gov.au/2035-emissions-reduction-targets.

40   "a direct threat": Adam Morton, "Australia urged not to rely on 'drug dealer's defence' for gas exports and help wean Japan off fossil fuels", *The Guardian*, 5 December 2023.

40   "phase out of fossil fuels", "Australia is not yet on track": Chris Bowen, Press conference at COP28, Dubai, Department of Climate Change, Energy, the Environment and Water, 2023.

43   "It's a bit like": Polly Hemming, "The government needs to stop using dodgy 'land use' accounting to suggest emissions are falling", The Australia Institute, 30 November 2023, accessed at australiainstitute.org.au/post/the-government-needs-to-stop-using-dodgy-land-use-accounting-to-suggest-emissions-are-falling.

43   other emissions only need to decline by 24 per cent: Climate Action Tracker, "Australia", 2023, accessed at climateactiontracker.org/countries/australia.

43–4   if you remove the inflated land-use figures: Hemming, "The government needs to stop".

44   only declining at around half the required rate: Climate Change Authority, 2023 *Annual Progress Report*, accessed at climatechangeauthority.gov.au/annual-progress-advice-0.

44   Australia is unlikely to get much below a 50 per cent reduction: John Quiggin, "Two charts in Australia's 2023 climate statement show we are way off track for net zero by 2050", *The Conversation*, 4 December 20023.

44   "People are getting carbon credits": Andrew Macintosh and Don Butler, "The unsafe Safeguard Mechanism: How carbon credits could blow up Australia's main climate policy", *The Conversation*, 10 November 2023.

45   around 80 per cent of pilot CCS projects: Stockholm Environment Institute, *Production Gap Report* 2023.

45   "We are now seeing": Global Carbon Capture and Storage Institute, *Global Status of CCS Report* 2021, accessed at globalccsinstitute.com/previous-reports.

46   global stocktake conducted at COP28: United Nations Framework Convention on Climate Change, "Outcome of the first global stocktake", 2023, accessed at unfccc.int/documents/636608.

46   "a near total phase-out": Stockholm Environment Institute, *Production Gap Report* 2023.

49   "The push for": Center for International Environmental Law, *Deep Trouble: The risks of offshore carbon capture and storage*, November 2023, accessed at ciel.org/

reports/deep-trouble-the-risks-of-offshore-carbon-capture-and-storage-november-2023.

51   fossil fuels in most countries are priced incorrectly: International Monetary Fund, *IMF Fossil Fuel Subsidies Data: 2023 Update*, 24 August 2023, accessed at imf.org/en/Publications/WP/Issues/2023/08/22/IMF-Fossil-Fuel-Subsidies-Data-2023-Update-537281.

51    $11.1 billion in subsidies to the fossil fuel industry: The Australia Institute, "Fossil fuel subsidies in Australia 2023", 2023, accessed at australiainstitute.org.au/report/fossil-fuel-subsidies-in-australia-2023.

52   produce up to 49 million tonnes: Climate Analytics, *Emissions Impossible: Unpacking CSIRO GISERA Beetaloo Middle Arm fossil gas emissions estimates*, accessed at climate analytics.org/publications/emissions-impossible.

53   could reduce carbon emissions 34 per cent: International Monetary Fund, *IMF Fossil Fuel Subsidies Data: 2023 Update*.

53   could raise revenue worth US$4.4 trillion in 2030: International Monetary Fund, *IMF Fossil Fuel Subsidies Data: 2023 Update*.

54   could deliver more than 80 per cent: International Energy Agency, *Net Zero Roadmap: A global pathway to keep the 1.5 °C goal in reach*, 2023 Update, IEA accessed at iea.org/reports/net-zero-roadmap-a-global-pathway-to-keep-the-15-0c-goal-in-reach.

54   "We are at a crossroads": Intergovernmental Panel on Climate Change, "The evidence is clear: the time for action is now. We can halve emissions by 2030", Press release, 2022, accessed at ipcc.ch/2022/04/04/ipcc-ar6-wgiii-pressrelease.

54–5 tripling of renewables would halve the need for coal power: Carbon Brief, "Why deals at COP28 to 'triple renewables' and 'double efficiency' are crucial for 1.5C", 2023, accessed at carbonbrief.org/qa-why-deals-at-cop28-to-triple-renewables-and-double-efficiency-are-crucial-for-1-5c.

57   "Consider the fact": Richard Denniss, National Press Club Address, The Australia Institute, 31 January 2024.

57–8 Garnaut and fellow economist Rod Sims: I.A. MacKenzie, "Ross Garnaut and Rod Sims have proposed a $100 billion-a-year fossil fuel tax – and it's a debate Australia should embrace", *The Conversation*, 15 February 2024.

58   Australia could halve its emissions by 2030: CSIRO, *Pathways to Net Zero Emissions – An Australian Perspective on Rapid Decarbonisation*, 2023, accessed at csiro.au/en/research/environmental-impacts/decarbonisation/pathways-for-Australia-report.

58   ultimately large-scale battery storage is a superior, more flexible alternative: Clean Energy Council, "Battery storage: The new, clean peaker", 2023, accessed at cleanenergycouncil.org.au/resources/resources-hub/battery-storage-the-new-clean-peaker.

59   electricity produced by renewables is cheaper: CSIRO, "The question of nuclear in Australia's energy sector", CSIRO, accessed at csiro.au/en/news/All/Articles/2023/December/Nuclear-explainer.

59   "The advice": Adam Morton, "Forget nuclear: Would Peter Dutton oppose a plan to cut bills and address the climate crisis?" *The Guardian*, 19 March 2024.

60   "The transformation": CSIRO, *Pathways to Net Zero Emissions*.

61   "It's devastating": G. Redfearn, "Fifth mass coral bleaching event in eight years hits Great Barrier Reef, marine park authority confirms", *The Guardian*, 8 March 2024.

61   two to four of these heating units can trigger the onset of bleaching: Terry Hughes, "The Great Barrier Reef's latest bout of bleaching is the fifth in eight summers – the corals now have almost no reprieve", *The Conversation*, 9 March 2024.

63   "We will not see": Joëlle Gergis, *Humanity's Moment: A climate scientist's case for hope*, Black Inc., Carlton, 2023.

Niki Savva

Lech Blaine has provided a thorough – if thoroughly depressing – dissection of the character and style of Peter Dutton. Despite this, Blaine's conclusion in *Bad Cop* is that none of Dutton's shortcomings or flaws renders him unelectable. Blaine is 100 per cent correct.

Despite some evidence which says Dutton is not a winner, including the loss of Aston and the inability to reclaim Dunkley, that does not mean the tactics he has employed to destroy Anthony Albanese are doomed to fail. Especially if Albanese allows it to happen. That is something else that comes through in Blaine's essay. Dutton is not the only one who steps up to the despatch box with a glass jaw. Albanese needs to fight smarter, not dirtier, and with greater discipline to avoid minority government or the shame of being a one-termer. He has spurts, then lapses.

As Blaine has shown, Dutton never takes his foot off Albanese's throat. He watched Tony Abbott do this to Kevin Rudd, then Julia Gillard, then Rudd again. Abbott boasted to me about this brutal but highly effective technique when he was Opposition leader. Abbott continues to tutor Dutton and to involve himself in depth in Liberal preselections in an effort to ensure the right (Right) candidates succeed. I call Abbott Terminator One and Dutton Terminator Two.

Dutton is both ruthless and relentless. Blaine has captured that side of him perfectly. It makes him dangerous. It also means he should not be underestimated. Blaine is frustrated by Dutton's ability to cast himself as closer to the working class despite his great friendship with Gina Rinehart and great wealth from canny property investments, but it shows how politically smart Dutton has been.

As one who has written – and said often – that Dutton is a much more engaging person privately, Blaine's essay forced me to stop and think if I had him wrong.

Yes and no.

Dutton retains a certain dark humour which journalists find appealing – I guess because it is impossible to survive in politics without it. Dutton gives in to that dark, funny side privately, while in public he fights a mighty battle to save himself and to save the Liberal Party from extinction in the face of threats to its left and right flanks. The teals and a more centrist Labor Party have chewed into the Liberals' seats, while One Nation and Clive Palmer have bitten into its conservative base.

Dutton's answer, as Blaine has shown, is to veer right, to go hard and to go dirty.

Niki Savva

Thomas Mayo

There are a multitude of reasons why I would not want to see Peter Dutton become Australian prime minister. Lech Blaine's essay brings them, and a few more, to light. But as an Indigenous Australian, I underline the 2023 referendum.

When I think about Dutton's choice to bind his front bench to the No campaign, I think about this passage from Noel Pearson's 2022 Boyer Lectures:

> If success in the forthcoming referendum is predicated on our popularity as a people, then it is doubtful we will succeed. It does not and will not take much to mobilise antipathy against Aboriginal people and to conjure the worst imaginings about us and the recognition we seek. For those who wish to oppose our recognition it will be like shooting fish in a barrel. An inane thing to do – but easy. A heartless thing to do – but easy.

Dutton has been around the traps. He would have known that few Australians know an Aboriginal and Torres Strait Islander person, and that they are therefore an easy target for tactics of confusion and fear-mongering.

Over the years, the antipathy to Indigenous Australians had settled at the bottom of the barrel. It was easy, but heartless, for an Opposition leader to stir it up.

It was painful to watch Peter Dutton and the Opposition front bench asking questions about the Voice to Parliament without accepting the answers – not from Indigenous leaders, nor from the most eminent legal authorities. Even more frustrating were his friends in the media who did the rest, parroting their queries as if he had uncovered a secret plot. "An inane thing to do – but easy."

It might be that we agree with arguments about flaws in the Yes campaign; we might speculate on what Prime Minister Albanese could have done differently; in hindsight, we might believe we should have waited for more favourable economic times. It is all fair game. But all analysis of what happened should start and end with Peter Dutton's decision to use a generous, heartfelt and modest proposal from the most disadvantaged and mistreated Australians as a political opportunity.

If we are to move forward, it is important that people are made aware of politicians who will try to take the "heartless – but easy" route to power. I hope we never reward such cruel cynicism from parliamentarians, especially not from those who aspire to be prime minister.

Lech Blaine has produced a very perceptive account of Peter Dutton, and it may be useful to Australians who, like me, fear he could be propelled to the role of Prime Minister of Australia at the next federal election. I will be buying more copies to circulate to as many swinging voters as possible between now and then. I'll be inviting them to do the same.

Thomas Mayo

Lachlan Harris

"Queenslander is not just a word, it's a call to action." So said Billy Moore, the legendary rugby league hardman who played seventeen State of Origin games in the 1990s. If you were watching the TV broadcast of Game 1 of the 1995 State of Origin series, you would have heard Billy steel the spine of his teammates by bellowing "Queenslander! Queenslander!" as he exited the tunnel onto the hostile turf of the Sydney Football Stadium. The Maroons, rank underdogs, went on to sweep the series. Billy Moore's heroics entered Origin folklore.

Reading Lech Blaine's deeply researched essay *Bad Cop*, it is clear there is quite a lot of Billy Moore energy in Peter Craig Dutton – the first leader of the Coalition to represent a Queensland electorate. Moore was skilful and relentless. So is Dutton. When biff was needed, Billy delivered biff. So does Peter Dutton. Off the field, Billy is an affable man. So is Peter Dutton. As his quote makes clear, Billy Moore believes the essential quality of a Queenslander is to be a man of action. As *Bad Cop* makes clear, so does Peter Dutton. This begs the question: will Peter Dutton, like Billy, channel his underdog energy into a famous victory? It's possible; in elections almost anything is. But the odds are stacked against him. Why? Because Billy Moore was playing rugby league, but Peter Dutton is practising politics.

The distinction matters, because in politics a relentless determination to act is frequently a flaw in the strategy of Opposition leaders. Hence it usually takes several election losses for Oppositions to learn that it's often the things you don't do that matter the most. The policies you don't launch. The fights you don't pick. The party shibboleths you don't pursue. Opposition leaders must show restraint, and restrain their party, to prove they are ready to govern. John Hewson, Mark Latham and Bill Shorten all learnt this lesson the hard way.

This does not mean that Oppositions are rewarded for doing nothing. On the contrary, it means good Opposition leaders refuse to allow themselves, or their party, to be drawn into every political debate that presents itself, rather seeking to

channel the voting public's attention into a small number of carefully curated political fault lines. Kevin Rudd used this strategy to focus the debate on WorkChoices
in 2007. Anthony Albanese mastered it in 2022. When Oppositions take every bait
presented, they let the people, or events, laying the bait determine the fault lines
they fight over. This matters because in close elections it is defining the questions,
not just knowing the answers, that usually determines who wins. In the '80s and
'90s, it took the Coalition five election losses to learn this lesson. Labor needed to
lose the next four to learn the same thing. Peter Dutton has not yet internalised
this truth. He may, but it often requires an election loss, or two, before you do.

It's important to note that abstention is not a prerequisite for victory if your
opponent is busy defeating themselves. Tony Abbott proved this in 2013. However,
the current Labor government is stacked with senior cabinet ministers who spent
the last decade mopping up the mess from Labor's last act of self-harm. Odds are
they've lost their taste for blood. When Opposition leaders design a political strategy that relies on major missteps by their opponent, they leave their adversary's
fate in their own hands. Oppositions can get lucky, but they are much more formidable when they plan to beat their opponent at their best. It is out of Peter
Dutton's control, but a sharp spike in interest rates, well above what is now priced
in, would also make the Albanese government very vulnerable to a mortgage-belt
revolt. It's a possible pathway to victory, but it's not a plan if it requires an act of
the political gods.

Before Labor supporters start popping champagne corks, they should consider
what *Bad Cop* makes clear. Namely, since entering parliament Peter Dutton has
assiduously developed a core political identity as a tough Queensland cop in a bad
world, who will do what needs to be done to keep us safe. It's a simple identity,
but in the right political conditions it's also very strong. If domestic, international
or border security fault lines define the next election, Peter Dutton will be hard to
beat. Labor must continue to do the work to ensure these issues are not central
to the campaign and pray to the political gods that an unforeseen event, like the
9/11 terror attacks, does not force the electorate's hand.

John Howard, Tony Abbott, Malcolm Turnbull and Scott Morrison all identified
Dutton's skill as a tool they could use to secure their own power. This is a serious
compliment to Mr Dutton. But as Lech Blaine suggests, there are warning signs
that Dutton may be much better at galvanising power inside the Coalition than he
is at delivering power to it. Over the past decade, as his star has risen, the Coalition's primary vote has plunged. At the 2013 election Dutton was a sidelined
shadow health minister – by 2016 he was a talismanic immigration minister. And
in 2019 and 2022 he was the Minister for Home Affairs, described by Blaine as

"perhaps the most powerful minister in Australian history." Over that period the Coalition's primary vote started at 45 per cent, dropped to 42 per cent in 2016, 41 per cent in 2019 and plunged to a record low of 35 per cent in 2022. Over a slightly longer period Labor's primary vote also declined, so it is clear Dutton is not the primary driver of this drop. But it still requires a bucketful of optimism to consider these trends and conclude that more Peter Dutton is the answer to the Coalition's woes.

Successful Opposition leaders are problem-solvers. Their rise solves the primary problem of an Opposition; namely, how to attract enough support to form government. But Dutton looks more like a paradox than a problem-solver to me. His appeal to the Coalition base is irresistible. Meaning his rise within the Coalition is unstoppable. But his appeal to a broad enough cross-section of voters to win an election appears chimeric. Meaning his pathway to victory is extremely narrow. This is particularly true given the rise of the teals as a centrist, non-Labor alternative, and the rebranding of the Greens as a generationally targeted renters' party. The machismo-fuelled refusal of the Liberal Party to preselect and promote female candidates and MPs makes this already complex challenge even more difficult. In a three- or four-cornered contest, one-dimensional political strategies become even more context-dependent for success. Meaning Dutton's action-oriented strategy becomes even more reliant on events, and outcomes, over which he has no control.

When Billy famously bayed "Queenslander! Queenslander!" he became a legend. Not because of his passion, but because his call to action worked. The Maroons won. It's too early to know if Peter Dutton will emulate Billy. But voters' preference for restraint, the stability of the Albanese government, the long list of issues crowding the political arena, changes to federal voting patterns, and the male-centric orientation of the Liberal Party all mean the odds of Peter Dutton becoming the next PM are long. What we do know is that Dutton's strategy faces the same test as Billy Moore's. Lauded if he wins. Ridiculed if he loses. Come to think of it, maybe rugby league and politics aren't that different after all.

Lachlan Harris

Mark Kenny

Labor government members were delighted as Christmas approached in 2009 when the hyper-aggressive Tony Abbott emerged from the Liberal scrum as the new leader of the Opposition. There was a similar sense of relief in progressive circles in the aftermath of Labor's win in 2022, when the only candidate for the Liberal Party leadership was Peter Dutton.

Both times, this optimism was hubris. Abbott's hammering shook Labor MPs into cutting down a popular first-term prime minister. And he was just getting started: the punchy Liberal then drove the government into minority in 2010 and stormed over the top in 2013. The effectiveness of Abbott's laser-like negativity and crisp repetitive messaging had been wildly underestimated.

When placed in this sobering historical context, Lech Blaine's illuminating essay on Dutton is timely and compelling. Both of these hardest-of-the-hard Liberal men had been described as unelectable. This was wrong the first time, and the jury is still out on the second.

When Dutton made his move on the moderate prime ministership of Malcolm Turnbull in 2018, the party room considered his selection unthinkable. "He would frighten voters away," I recall one minister telling me. The beige and slippery Scott Morrison was the compromise.

But times change.

With the art of a writer and the sensibility of a Queenslander, Blaine addresses the Dutton question, his current adjacency to power and his grim determination to become this nation's next prime minister – as early as 2025. It is a trajectory that takes us from scoffing impossibility to surprising plausibility.

This is not a sympathetic account, but neither is it, as some apologists for the conservative cause have depicted it, unflinchingly hostile. Although its opening lines can certainly be read that way: "Peter Dutton eats bleeding-heart lefties for breakfast. He is tall and bald, with a resting death stare. His eyes – two brown beads — see evil so that the weak can be blind. His lips are allergic to political correctness."

Rather, Blaine is making Dutton's progressive framing explicit so as to fill out who his subject is, and why he may prevail in his mission to commune with enough voters to lead an Australian conservative government.

Many on the left still think this impossible. But do they watch Nine's *Today* or Seven's *Sunrise*? For Blaine, and for this writer too, Dutton is both more ruthless than Abbott and more dexterous, more intellectually organised, and yet in some important senses more unremittingly bleak in his worldview.

Singling him out from the cheery pattern of Liberal predecessors — old boys from elite private schools and top universities whose progression has been marked by open doors, privileged access and consistent promotions — Blaine notes that Dutton was no natural student, did not enjoy university life and "bombed out" to become a Queensland copper. There he learnt about depravity, crime and the collapse of social order when rules and prohibitions are not sufficiently corrective.

Blaine characterises Dutton's overall strategy in Trumpian terms as "make Australia afraid again" and says Dutton isn't as happy-go-lucky as other Liberal leaders. "He views the world with the pessimism of a Russian novelist."

It feels like an incongruous observation to make of one so defiantly suburban. Perhaps, then, "Dutts," as his colleagues call him, is more of a piece with his immediate predecessor, the daggy suburban dad, Scott Morrison? Yes and no. Dutton, Blaine reminds us, proudly told Niki Savva that he had voted for same-sex marriage. This presumably was a declaration of difference from the former's Pentecostal conservatism.

This wasn't quite true, but nor was it completely false. Dutton had actually opposed same-sex marriage during the public debate for the 2017 postal survey — itself a creative "fix" he had cleverly conceived to extract the government from its ideological cul de sac — but had then voted for it in parliament, in line with the strong majority view in his electorate. Was it the sacred principle of faithful democratic representation that straightened him out in the end, or just the glimpse of electoral annihilation?

It's worth remembering that while Abbott led the national case against same-sex marriage, his own electorate of Warringah returned the second-highest Yes vote among Liberal-held seats in New South Wales. Abbott was turfed out of parliament at the next election — Dutton wasn't.

The important thing to understand about Dutton, then, is his willingness to learn and adapt, but only to the minimum degree necessary to neutralise a problem while retaining his basic outline.

The limits of this were evident in his evaluation of Morrison's loss of six inner-urban Liberal heartland jewels – Wentworth, North Sydney, Mackellar, Kooyong, Goldstein and Curtin – to teal independents in 2022.

Dutton's response to the gutting of his party was telling. As the new leader, he chose to double down on the city versus suburban divide he sees emerging in Australian society. He wants to position the Coalition firmly in the latter camp so that it can campaign "unshackled" by the pragmatism and dual messaging required to speak to both educated, cosmopolitan Australia and the rest.

Blaine lays out the difficult maths of this strategy but can't find a convincing method of making an Australia bifurcated in this way add up politically. At times, he seems to suggest that Dutton – a known hardliner on immigration – is happy just to win the policy fight even if the gleaming office is denied him.

Dutton doesn't need to become prime minister to redraw the battlelines of Australian politics, Blaine writes, arguing that "his fight with Albanese over the suburbs and regions was aways going to drag the political conversation rightwards: on race, immigration, gender and the pace of transition away from fossil fuels."

While it is observably true that Oppositions can shape policy and the positioning of governments – as Abbott did to Kevin Rudd and Whitlam did to McMahon – Dutton is no less focused on the prize of office than any of his predecessors. Not for him the ambivalence rightly or wrongly ascribed to Kim Beazley and Andrew Peacock.

That said, the task before Dutton would be arithmetically challenging even if he had not chosen what looks to be the harder of two paths back to power.

He could have acknowledged the failings that had sent lifelong Liberals away from the party in its leafy heartland and promised to address them (women's representation, endless climate denialism, political corruption and lack of transparency), but instead he has decided to leave these voters to the Greens, Labor and teals.

Meanwhile, he seeks to turn the outer suburbs and regions blue. But should he really dismiss such erstwhile Liberal voters as unworthy of respect? Was this a "basket of deplorables" mistake, the significance of which commentators largely missed? Or is his new Liberal Party of workers, evoking the old Reagan Democrats idea, a stroke of electoral genius?

Whenever I ask Labor strategists about this, they challenge me to name the Labor seats Dutton hopes to turn Liberal. It is a fair question, seeing as he needs about fifteen of them. He may sway a significant number of blue-collar voters and still not pick up a single Labor seat. That would be that for his strategy. And for him.

This is the challenge for Dutton, and it is one he has already failed twice. In 2023, he became the first Opposition leader to cede a seat (Aston) to a sitting government at a federal by-election in 102 years. More tellingly again, this year he failed to wrest federal Dunkley from Labor during a cost-of-living crisis.

Blaine puts some of this down to a fatal mismatch between Dutton's rhetoric and his party's unchanged protection of wealth. He says Dutton's "tunnel vision" on representing the battlers in the suburbs and regions sits oddly with the Liberal Party's commitment to retaining tax loopholes that disproportionately benefit the kinds of Australians who live in what are now "teal" electorates.

Ditto for the promised Stage 3 tax cuts that were going to hand more than $9000 annually to the well-off. Labor, to its credit, halved them anyway.

It is an open question whether Dutton can find a way through this structural contradiction and, even if he can, whether voters will notice.

The Australia that Dutton wants to valorise and thus mobilise – some configuration of Menzies' forgotten Australians, Howard's battlers and Abbott's tradies – may no longer exist in sufficient numbers or concentrations to be decisive. "The working class of the twenty-first century is more feminine and ethnically diverse. And they are much less socially mobile, thanks to the deterioration of the Australian dream," Blaine concludes.

For current aspirants to that dream, the situation has become impossible, and that, in the end, might be enough to do in Dutton's illusory promise. You cannot complain about barriers to economic participation while refusing even to discuss their removal.

And if Dutton's pitch is to be more to the heart than the wallet, he has another problem. Australia has changed. It is more feminised and diverse, and identity, whether he likes it or not, is more fluid. Even his own electors know that. Again, you cannot abuse your way to power. At least I hope not.

For true small-L liberals, this may be a social reality to which some political adaptation is possible, but for a famously anti-woke, hardline ex-Queensland copper, well, it's just not acceptable, is it?

Mark Kenny

Robert Wood

In Lech Blaine's Quarterly Essay about Peter Dutton, *Bad Cop*, he maps out the right-winger's Suburban Strategy and identifies something true about Australian democracy. This is the important fact that suburbia, inclusive of exurban places further away from the city and towards the regions, is where elections are won and lost. Quite simply, the suburbs are where Australia takes place, from our politics to our culture, something that the mid-century intellectuals Robin Boyd, Patrick White and Donald Horne understood and often railed against. I have no doubt that Dutton's plan instrumentalises this part of the nation for his own ideological ends, and that should have us concerned.

I also think Blaine is right to suggest this Suburban Strategy is a two-election attempt to redraw the map and will be successful eventually: Albanese will likely govern in minority in 2025 and lose in 2028. The current response, based on the last federal election, often boxes itself in, making itself relevant only within the latte line. The inner city can be an echo chamber even as it has fragmented along Green, Labor and teal lines. It often fails to speak beyond internalised borders, beyond where the trams stop, where the wine is natural and the coffee cold-brewed. It forgets where "the people" actually live and where the votes are. Hawke with his larrikins, Howard with his battlers and Rudd with his ordinary working families all got this in an instinctive way. Dutton is right that he needs a Suburban Strategy, even as he is pulled ever rightwards by his regional flank, which panders to an outdated idea of what country Australia wants. Progressive intellectuals and party strategists need to consider what the suburbs truly want and, of course, what exurban and country areas consider to be important, over and above the clichés.

Thinking through the changes in country Australia matters here too as an ideological and practical reality, with forerunning independents Tony Windsor, Rob Oakeshott and Cathy McGowan pointing towards something significant. That is a type of "agrarian socialism" that challenges the Nationals' hegemony, a kind of thinking and lifestyle politics that has given rise to the Lock the Gate movement as farmers support action on climate change and oppose multinational mining; and that finds expression in support for refugees from Biloela to Katanning, most especially with the Murugappan family. So although elections are decided in the suburbs, the national political discourse needs progressives to take country areas seriously. This is precisely because of the romanticisation of "the bush" for the suburban imagination and a Nationals Party that is increasingly out of touch. Rather than a three-way fight from Curtin to Kooyong to Marrickville, it's important to have a proper debate in New England and Forrest.

What is needed, then, is not Dutton's anxiety, fear and dog-whistling, but the recognition of the ordinary decency of average people by democratic representatives and our intellectuals. To speak to the suburban voter who wants to live up to the promise of what our politics can be. This is a politics grounded in tolerance, social democracy and sovereignty, all values that Australians live by (though we call them harmony, fairness and truth). All of us would do well to lift our gaze from the mundanity of mudslinging and begin to articulate a set of coordinated policies that build a true collaboration among progressive forces, one that is strategically minded enough to include the suburbs, rural and remote areas.

If Dutton's approach is indeed based on the Republicans' Southern Strategy, as Blaine suggests, then we might also recall the contrasting wave of support Barack Obama generated in 2008 in places such as Virginia, North Carolina and Florida. When he won, Obama said, "There are no red states, no blue states, just the United States of America." Here we might need an election or two in which progressives proclaim there is no inner city, no outer regions, just the suburbs of Australia.

Robert Wood

Paul Strangio

Reading Lech Blaine's eviscerating and disquieting portrait of Peter Dutton, *Bad Cop*, my mind drifted, perhaps counterintuitively, to parallels between Dutton and Bill Hayden. Hayden, who passed away last October, was another former police officer-turned-politician from the Sunshine State, member for another sprawling outer-suburban Brisbane seat, and another federal Opposition leader. Like Dutton, he was exposed to the dark underbelly of humanity as a young Queensland copper.

There, however, the similarities end. While Dutton's policing days blackened his heart, fuelling an urge to punish the troubled and dispossessed in life, Hayden concluded that the best solution to the social problems he encountered on the beat was to uplift the vulnerable through the redemptive power of government. While Hayden's humble working-class origins inclined him towards collectivist egalitarianism, Dutton is a model of sharp-elbowed individualistic aspiration, encapsulated in his own rise from the lower-middle-class mortgage belt to riches via property investment. While Hayden, in defiance of the historical racial prejudices of his home state and inspired by another ex-copper cum politician, Jim Cairns, was among the early Labor converts to opposition to the White Australia policy, Dutton has a history of playing with the fire of race-based prejudice. While Hayden as minister for social security in the Whitlam government introduced a landmark extension of the welfare state, the single mothers' benefit, and fathered a pioneering universal health insurance scheme, Medibank, the forerunner of Medicare, Dutton as minister for workplace relations in the Howard Coalition government earned prime-ministerial favour by cutting welfare payments and hunting dole "cheats." Later, as health minister in the Abbott Coalition government, he presided over the slashing of spending in the portfolio and an unsuccessful attempt to apply a GP co-payment that would have undermined Medicare's core principles. And while, as Opposition leader between 1977 and 1983, Hayden spearheaded a major renovation of Labor's policy program to restore the party's shattered electoral credibility following the upheavals of the Whitlam years and assembled around him a formidable shadow ministerial team, Dutton has little to show in the way of policy or party rejuvenation since becoming Liberal leader in 2022. While Hayden's expansive legacy as Labor leader laid the groundwork for the Hawke/Keating reform era, marking him out as possibly Australia's finest Opposition leader who never became prime minister, Dutton's mission in Opposition appears aimed at debauching the national political conversation, and about sidling into office by frightening voters into submission.

So, despite the similarities in their back stories, the differences between Hayden and Dutton could hardly be starker. Arguably, the contrast is a disturbing marker of the degeneration of the political class across generations, of the retreat from a milieu of enlightened social-democratic optimism to irrational conservative populist pessimism, and of the decline of a political sensibility of compassion and empathy to one of stony-heartedness.

Though Dutton emerges from the pages of Blaine's essay as an uber version of the political strongman, he is not sui generis. Most obviously, authoritarian conservative populism is a worldwide trend, but Blaine is silent on international comparisons, apart from passing references to Trumpism and, from the past, Richard Nixon, whose invocation of the "silent majority" was abidingly influential for conservative politics in the United States and beyond. Political psychology literature has long recognised Dutton's leadership type. One of the most fertile and original writers in the field, Australia's Graham Little, delineated three leadership types: the strong leader, the inspirational leader and the group leader. Strong leaders have been most prevalent on the Coalition side of politics. John Howard, Tony Abbott and now Dutton are recent cases in point. On the other hand, Labor leaders are more likely to fall into the inspirational or group leader categories. Anthony Albanese, with his promise of a "cuddlier, less bloodthirsty" politics, as Blaine observes, his collegial approach to governing and promotion of community solidarity, fits readily into the latter camp. What particularly defines the strong leader is their trading in fear and insecurity. They project the world as a menacing place, with competition the primary motor force of human relations. These are hallmarks of Dutton's politics. The challenge for the strong leader is to conjure up and orchestrate community anxieties, to identify threats and to establish themselves as a decisive counter-agent to those threats. A key question lingering over Dutton's leadership project is whether at this moment Australians feel sufficient trepidation about the international and domestic environment to be susceptible to his Hobbesian world view.

Blaine usefully, and I think credibly, places Howard and Dutton on a Liberal Party leadership continuum, although Abbott's place in that evolution is largely overlooked. In one of his many arresting turns of phrase, Blaine writes: "The leadership of the Liberal Party can be broken into three historical epochs ... The Highbrow Liberals, led by Menzies. The Middlebrow Liberals, led by Howard. And eventually the Lowbrow Liberals, led by Peter Dutton." The essay helps remind us that Howard, whose record in office is conventionally airbrushed to enable his sanctification as a benign and wise elder statesman rather than the architect of many of the nation's ongoing ills (including the distortions in housing policy that have recently come into sharp relief), was responsible for the Liberals' movement towards conservative populism. Unexceptionally, Blaine argues that Dutton is propelling the party further along a reactionary path, "with much less emphasis on economics and much less subtlety on race relations" than Howard practised.

Another distinction between the two, which is insufficiently probed by Blaine, lies in their psychologies. Howard was utterly comfortable in his own skin – I once pondered if Australia had ever had a national leader more psychologically bulletproof – and his fear-mongering was above all motivated by political ends. There was little hint that this was a man attending to his own deep-seated insecurities. However, Blaine senses that below the surface there is a fragility to Dutton, concluding that although he is "tall and strong at first glance ... when you watch him for a long time, you can see that the man is small and scared." In other words, Dutton's fostering of social anxieties can be attributed, at least partially, to his own inner fears. It is an intriguing notion. One of the fundamental insights of political psychology is that power is pursued as a means of compensation for subterranean feelings of lack of self-worth. Events in Dutton's early life outlined by Blaine suggest some causes for insecurity as well as resentment – his parents' marriage failing, his own first marriage ending after just five months, his flunking out of university, and his doomed tilt at the Queensland parliament at age nineteen. Yet Blaine does not draw a compelling connection between these setbacks and Dutton's political outlook. He leaves cryptic his conclusion that the Liberal leader "is small and scared." This idea of an inherent vulnerability to Dutton is also not squared with other aspects of the Opposition leader noted in the essay – for example, the colleagues attesting to his steadiness of character and absence of self-aggrandisement. And there is the elephant-like hide he displays towards critics, especially those he disdains as politically correct elites. Blaine memorably quotes Dutton declaring, "They [his left-wing critics] don't realise how completely dead they are to me." These strains in Blaine's analysis leave the suspicion that his peroration on a shrunken and weak Dutton springs as much as anything from a distaste for his subject, a distaste that he struggles to disguise.

A nub of Blaine's essay is the question of whether Dutton's strategy for primeministerial power is viable. The heart of this strategy is for the Coalition to focus on winning outer-suburban, rural and regional seats, even at the expense of forsaking formerly safe Liberal electorates in affluent inner-city suburbs snatched by teal independents and which, to lesser degree, are vulnerable to encroachment from the Greens and Labor. This would accentuate a refashioning of the party's appeal begun under Howard, although Howard successfully managed to build an electoral coalition between suburban aspirational workers (the "Howard battlers") and voters in the Liberals' wealthy traditional blue-ribbon seats. In the words of another of the party's hard-right Queensland MPs, Matt Canavan, the Liberals must embrace "boganisation," transforming their party into the natural home for non-tertiary-educated, ordinary, working suburbanites. According to Blaine, for Dutton's strategy to succeed it "would take a history-making realignment, no less seismic in nature than Menzies' recruitment of Catholics, Whitlam's recruitment of graduates and Howard's recruitment of blue-collar battlers."

Blaine cites electoral and demographic evidence, as well as the voices of dissenting Liberal insiders, that cast doubt on Dutton's strategy. He reasonably asks if Dutton is "leading the Liberal Party into an electoral cul-de-sac." Examples of that evidence include that Labor holds a preponderance of outer-metropolitan seats, the majority of these by hefty margins; that many of the outer-suburban electorates of Sydney and Melbourne, unlike Dutton's own relatively monocultural Brisbane seat of Dickson, are home to significant Asian and Middle Eastern communities that are likely to be alienated by his racial dog-whistling (and we can add the Liberals' dogmatic pro-Israel position on the Gaza conflict); and, more generally, signs that younger generations of voters, unlike older cohorts, are not evolving towards conservatism as they age. Notwithstanding such evidence, Blaine hedges on whether Dutton is capable of fulfilling his prime-ministerial ambition. He floats the possibility of Dutton succeeding with a two-term strategy in which Labor is reduced to minority government in 2025 (a real chance on current polling on the ALP's primary vote), thereby forcing Albanese into an alliance with the Greens that would heighten the chance of the Liberals reclaiming the teal seats at the following election. It is a scenario that Blaine contends makes a Dutton-led electoral realignment "less of a pipedream." One thing he does not consider is that in the context of the eroding primary vote support for both the Labor and Liberal parties, the Coalition is equally likely to find itself without a majority when it next has the opportunity to form office. On the face of it, Dutton's strongman leadership style would be ill-suited to the power-sharing, negotiation and compromise required in a minority government.

Naturally, Blaine devotes attention to the Voice referendum, with Dutton's No campaign open to being read as a dummy run for his outer-suburban-focused electoral push. That Dutton was rehearsing that strategy is abundantly clear in statements he made during the campaign, such as, "If you're living in an outer-suburban area, the new Labor Party is not for you, it's for inner-city trendies" and "The Liberal Party – the Coalition – is today's party for the worker." The fault lines apparent in the referendum result provided grist for Dutton's electoral realignment mill. Yes voters were more likely to be from affluent inner-city areas, tertiary-educated and young, while No voters were more likely to live in outer suburbs and regional and rural areas, to have lower levels of education, and be older and less well off. It suggested, as did the 1999 republic referendum, that there are two Australias: one cosmopolitan, confident and progressive, the other provincial, apprehensive and conservative.

If Dutton obtained reinforcement for the direction he is charting for the Liberal Party from the No victory in the Voice referendum, it is nevertheless doubtful that the patterns evident in that result will translate to a broader election contest. Traditional working-class Labor seats that voted No in the 1999 republic referendum did not defect from the ALP at subsequent elections. To the contrary, the erosion of Labor's primary vote in the twenty-first century has chiefly involved inner-city progressives transferring their allegiance to the Greens. This is not to say Labor can be complacent about retaining the loyalty of workers in the outer suburbs – there were portents of a working-class rebellion against the party in parts of Melbourne at the 2022 election. But these voters mostly hived off to a melange of minor parties rather than to the Liberals. Unquestionably, as the bases of both major parties continue to crumble in a landscape of heightened social diversity and complexity, there will be growing opportunities for each to hijack, even if temporarily, segments of the other's traditional supporters. Yet a wholesale and permanent electoral realignment, with the Liberals becoming the "party for the worker," as is Dutton's aspiration, is a remote prospect any time soon.

Moreover, in addition to some of the evidence supplied by Blaine, other factors render Dutton's muscular conservatism a dubious fit for contemporary Australia. One is the electoral influence of women, particularly the fast-growing number of university-educated female voters. Dutton's highly charged, aggressively masculinist leadership risks repelling rather than attracting this cohort. For some time now, women have been abandoning their habit of voting conservative, a trend Dutton appears singularly unsuited to arrest. Perhaps most fundamentally, I suspect that Dutton has misread the national psyche. Despite significant social strains and uncertainties in the international and domestic climate, Australia is more optimistic, forward-looking and generous-hearted than he banks on. It is less scared, less paranoid. And we can also find comfort in history as to why Dutton will most likely fail in his quest for prime-ministerial power. Since World War II, no first-cab-off-the-rank Opposition leader (the individual who led a major party immediately after it lost government) has made it to the Lodge. Instead, their destiny has been to fall by the wayside. Considering the loud warning bells rung by Blaine's valuable and chilling examination of Dutton, we ought to be relieved that political redundancy is his probable fate.

Paul Strangio

Lech Blaine

I was pleased to read such deeply engaged responses to *Bad Cop*. Lachlan Harris shines a spotlight on the best exemplification of the Queensland psyche: State of Origin rugby league. I shied away from mentioning rugby league in my essay, given that I have perhaps been over-reliant on rugby league analogies in my past writing, and Peter Dutton doesn't publicly associate himself with the sport in the same way that Scott Morrison did.

Niki Savva's response remarked upon a more likeable private side to Dutton. In combination with Harris's football-themed response, it made me think about an obvious analogy for Dutton in the sporting world: famed rugby league coach Wayne Bennett. Like Dutton, Bennett is a former Queensland police officer. He has built an illustrious résumé from a deep siege mentality. Bennett belittles journalists, or barely engages with them. And yet – like Dutton – people who have worked with Bennett closely insist there is a much warmer and more humorous private side to his often-irritable public persona.

Bennett's willingness to attack the media kindles an intense loyalty in his players and supporters. In political parlance, he is a master at "throwing red meat to the base." It has helped make him arguably the greatest Australian sporting coach of all time. But such an approach has its limits in a political context. For a start, Bennett's main audience is a small group of hypermasculine men. His job is to inspire devotion in them, not to persuade supporters of rival teams to switch their allegiance. Whereas that is the job of Opposition leaders.

Dutton pressingly needs to address the Coalition's electoral problem with women, which was probed by Paul Strangio in his response to the essay. Lachlan Harris implies that key to being an Opposition leader is the capacity to disappoint your most diehard supporters. This won't come naturally to Dutton, for emotional and practical reasons. He is a diehard of the Liberal Party. His rise to the top of the party has been aided and safeguarded by partisan commentators who won't tolerate compromise.

Paul Strangio's response gave me a lot to ponder. Bill Hayden provides a fascinating counter-study to Dutton. Space didn't allow me to explore this, apart from a passing reference. For much the same reason, I didn't delve into international examples of "strongman" leaders and their growing prominence.

Strangio is entitled to question my conclusion about Dutton's inner anxieties. I am not a psychologist. I drew much from Dutton's admission that he probably sustained PTSD in the police force, and the way that those experiences continue to echo in his political decisions. It is hard to be diagnosed with clinical PTSD. I have suffered significant trauma, and yet don't suffer from PTSD. If we are to take Dutton at his word, I don't think it is a stretch to imagine his inner world, or the relationship between that private space and his public actions. Strangio is right that I might have established these connections more clearly in the essay.

I agree with both Niki Savva and Mark Kenny that Dutton himself isn't incontrovertibly "unelectable," despite the structural issues facing the Coalition. Dutton is loathed by cosmopolitan political junkies, but is still a mystery to disengaged voters, and even to many rusted-on conservatives. Political junkies overestimate how forensically the public remembers the various controversies of Dutton's career. Although Dutton may overestimate this too. His actions and rhetoric after becoming Opposition leader suggest that he didn't believe a more thoroughgoing attempt to recast his public image was possible.

I am glad Mark Kenny didn't find the essay "unflinchingly hostile," as some on the right have interpreted it as being. At the same time, some progressive readers have found it too sympathetic. These differing interpretations are grist for the mill of an essayist. It is a privilege to be criticised and disagreed with. But I think the inability of many conservatives to see that I've taken Dutton seriously points towards a wider siloing of public debate in Australia. The accusation of bias is frequently used to dismiss perspectives that don't come from natural allies. Political movements that only listen to hagiographers are doomed for irrelevance, both on the left and, in this case, on the right.

Thomas Mayo is right to highlight the stifling of debate following the Voice referendum. There was an air of embarrassment in Australia's media and political classes that they had been so out of step with the eventual result. Critique of the Yes campaign – and of Albanese's timing for the referendum – is essential. So is analysis of the divide between university-educated and non-university-educated voters, which is often misunderstood or dismissed by progressives. So too the divide between inner-city and outer-suburban voters, which was emphasised by Robert Wood in his response to the essay. These fissures exist and can't be ignored.

But the fact Dutton was on the winning side of the referendum doesn't let him off the hook. He is the alternative prime minister of Australia, not a single-issue campaigner. His attacks on the Yes campaign went beyond merely expressing scepticism about the legal mechanism of the Voice. He went for the jugular of reconciliation. It is necessary to critique his actions and rhetoric during the campaign, and measure them against the future policy offerings of the Coalition. This goes to Dutton's insistence that he is a more open-minded politician than the one who boycotted the Apology. So far, there is little evidence of this.

Lech Blaine

Lech Blaine is the author of the memoir *Car Crash* and the Quarterly Essays *Top Blokes* and *Bad Cop*. He was the 2023 Charles Perkins Centre writer-in-residence. His writing has appeared in *Good Weekend, Griffith Review, The Guardian* and *The Monthly*.

Joëlle Gergis is an award-winning climate scientist and writer. She served as a lead author for the IPCC's Sixth *Assessment Report* and is the author of *Humanity's Moment: A climate scientist's case for hope* and *Sunburnt Country: The history and future of climate change in Australia*. Joëlle also contributed chapters to *The Climate Book* by Greta Thunberg and *Not Too Late: Changing the climate story from despair to possibility*, edited by Rebecca Solnit and Thelma Young Lutunatabua.

Lachlan Harris is the co-founder of One Big Switch, Budgy Smuggler and the Bondi to Manly Walk. He was a political staffer for the Australian Labor Party from 2003 to 2010.

Mark Kenny was national affairs editor for *The Age* and the *The Sydney Morning Herald* and is a professor at the Australian Studies Institute at the Australian National University.

Thomas Mayo is a Torres Strait Islander man born on Larrakia country in Darwin. He is the author of *Finding the Heart of the Nation, Dear Son* and the children's books *Finding Our Heart* and *Freedom Day*, and co-author of *The Voice to Parliament Handbook*.

Niki Savva is a journalist, author and former senior adviser to Prime Minister John Howard and Treasurer Peter Costello. Her books include *Bulldozed, Plots and Prayers* and *The Road to Ruin*.

Paul Strangio is emeritus professor of politics in the School of Social Sciences at Monash University and the author or editor of a dozen books on Australian politics. One of his current projects is a study of Australia's best prime ministers.

Robert Wood is director of writing and publishing at the Centre for Stories, an arts non-profit that uses stories for social cohesion. He is a Sir Edward "Weary" Dunlop Fellow with Asialink and the University of Melbourne in 2024.

WANT THE LATEST FROM QUARTERLY ESSAY?

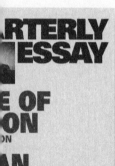

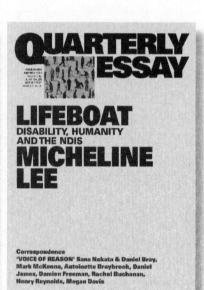

LIFEBOAT
DISABILITY, HUMANITY AND THE NDIS
MICHELINE LEE

Correspondence
'VOICE OF REASON' Sana Nakata & Daniel Bray,
Mark McKenna, Antoinette Braybrook, Daniel
James, Damien Freeman, Rachel Buchanan,
Henry Reynolds, Megan Davis

THE GREAT DIVIDE
AUSTRALIA'S HOUSING MESS AND HOW TO FIX IT
ALAN KOHLER

Correspondence
'LIFEBOAT' Bill Shorten, Monique Ryan,
Sam Drummond, Rhonda Galbally, Sam Bennett,
Robbi Williams, Carly Findlay, Micheline Lee

Subscribe to the Friends of Quarterly Essay email newsletter to share in news, updates, events and special offers.

quarterlyessay.com.au/signup

QUARTERLY ESSAY
BACK ISSUES

- [ ] **QE 1** ($27.99) Robert Manne *In Denial*
- [ ] **QE 2** ($27.99) John Birmingham *Appeasing Jakarta*
- [ ] **QE 3** ($27.99) Guy Rundle *The Opportunist*
- [ ] **QE 4** ($27.99) Don Watson *Rabbit Syndrome*
- [ ] **QE 5** ($27.99) Mungo MacCallum *Girt By Sea*
- [ ] **QE 6** ($27.99) John Button *Beyond Belief*
- [ ] **QE 7** ($27.99) John Martinkus *Paradise Betrayed*
- **QE 8** Amanda Lohrey *Groundswell* OUT OF STOCK
- [ ] **QE 9** ($27.99) Tim Flannery *Beautiful Lies*
- [ ] **QE 10** ($27.99) Gideon Haigh *Bad Company*
- [ ] **QE 11** ($27.99) Germaine Greer *Whitefella Jump Up*
- [ ] **QE 12** ($27.99) David Malouf *Made in England*
- [ ] **QE 13** ($27.99) Robert Manne with David Corlett *Sending Them Home*
- [ ] **QE 14** ($27.99) Paul McGeough *Mission Impossible*
- [ ] **QE 15** ($27.99) Margaret Simons *Latham's World*
- [ ] **QE 16** ($27.99) Raimond Gaita *Breach of Trust*
- [ ] **QE 17** ($27.99) John Hirst *'Kangaroo Court'*
- [ ] **QE 18** ($27.99) Gail Bell *The Worried Well*
- [ ] **QE 19** ($27.99) Judith Brett *Relaxed & Comfortable*
- [ ] **QE 20** ($27.99) John Birmingham *A Time for War*
- [ ] **QE 21** ($27.99) Clive Hamilton *What's Left?*
- [ ] **QE 22** ($27.99) Amanda Lohrey *Voting for Jesus*
- [ ] **QE 23** ($27.99) Inga Clendinnen *The History Question*
- [ ] **QE 24** ($27.99) Robyn Davidson *No Fixed Address*
- [ ] **QE 25** ($27.99) Peter Hartcher *Bipolar Nation*
- [ ] **QE 26** ($27.99) David Marr *His Master's Voice*
- [ ] **QE 27** ($27.99) Ian Lowe *Reaction Time*
- [ ] **QE 28** ($27.99) Judith Brett *Exit Right*
- [ ] **QE 29** ($27.99) Anne Manne *Love & Money*
- [ ] **QE 30** ($27.99) Paul Toohey *Last Drinks*
- [ ] **QE 31** ($27.99) Tim Flannery *Now or Never*
- [ ] **QE 32** ($27.99) Kate Jennings *American Revolution*
- [ ] **QE 33** ($27.99) Guy Pearse *Quarry Vision*
- [ ] **QE 34** ($27.99) Annabel Crabb *Stop at Nothing*
- [ ] **QE 35** ($27.99) Noel Pearson *Radical Hope*
- [ ] **QE 36** ($27.99) Mungo MacCallum *Australian Story*
- [ ] **QE 37** ($27.99) Waleed Aly *What's Right?*
- [ ] **QE 38** ($27.99) David Marr *Power Trip*
- [ ] **QE 39** ($27.99) Hugh White *Power Shift*
- [ ] **QE 40** ($27.99) George Megalogenis *Trivial Pursuit*
- [ ] **QE 41** ($27.99) David Malouf *The Happy Life*
- [ ] **QE 42** ($27.99) Judith Brett *Fair Share*
- [ ] **QE 43** ($27.99) Robert Manne *Bad News*
- [ ] **QE 44** ($27.99) Andrew Charlton *Man-Made World*
- [ ] **QE 45** ($27.99) Anna Krien *Us and Them*
- [ ] **QE 46** ($27.99) Laura Tingle *Great Expectations*
- [ ] **QE 47** ($27.99) David Marr *Political Animal*
- [ ] **QE 48** ($27.99) Tim Flannery *After the Future*

- [ ] **QE 49** ($27.99) Mark Latham *Not Dead Yet*
- [ ] **QE 50** ($27.99) Anna Goldsworthy *Unfinished Business*
- [ ] **QE 51** ($27.99) David Marr *The Prince*
- [ ] **QE 52** ($27.99) Linda Jaivin *Found in Translation*
- [ ] **QE 53** ($27.99) Paul Toohey *That Sinking Feeling*
- [ ] **QE 54** ($27.99) Andrew Charlton *Dragon's Tail*
- [ ] **QE 55** ($27.99) Noel Pearson *A Rightful Place*
- [ ] **QE 56** ($27.99) Guy Rundle *Clivosaurus*
- [ ] **QE 57** ($27.99) Karen Hitchcock *Dear Life*
- [ ] **QE 58** ($27.99) David Kilcullen *Blood Year*
- [ ] **QE 59** ($27.99) David Marr *Faction Man*
- [ ] **QE 60** ($27.99) Laura Tingle *Political Amnesia*
- [ ] **QE 61** ($27.99) George Megalogenis *Balancing Act*
- [ ] **QE 62** ($27.99) James Brown *Firing Line*
- [ ] **QE 63** ($27.99) Don Watson *Enemy Within*
- [ ] **QE 64** ($27.99) Stan Grant *The Australian Dream*
- [ ] **QE 65** ($27.99) David Marr *The White Queen*
- [ ] **QE 66** ($27.99) Anna Krien *The Long Goodbye*
- [ ] **QE 67** ($27.99) Benjamin Law *Moral Panic 101*
- [ ] **QE 68** ($27.99) Hugh White *Without America*
- [ ] **QE 69** ($27.99) Mark McKenna *Moment of Truth*
- [ ] **QE 70** ($27.99) Richard Denniss *Dead Right*
- [ ] **QE 71** ($27.99) Laura Tingle *Follow the Leader*
- [ ] **QE 72** ($27.99) Sebastian Smee *Net Loss*
- [ ] **QE 73** ($27.99) Rebecca Huntley *Australia Fair*
- [ ] **QE 74** ($27.99) Erik Jensen *The Prosperity Gospel*
- [ ] **QE 75** ($27.99) Annabel Crabb *Men at Work*
- [ ] **QE 76** ($27.99) Peter Hartcher *Red Flag*
- [ ] **QE 77** ($27.99) Margaret Simons *Cry Me a River*
- [ ] **QE 78** ($27.99) Judith Brett *The Coal Curse*
- [ ] **QE 79** ($27.99) Katharine Murphy *The End of Certainty*
- [ ] **QE 80** ($27.99) Laura Tingle *The High Road*
- [ ] **QE 81** ($27.99) Alan Finkel *Getting to Zero*
- [ ] **QE 82** ($27.99) George Megalogenis *Exit Strategy*
- [ ] **QE 83** ($27.99) Lech Blaine *Top Blokes*
- [ ] **QE 84** ($27.99) Jess Hill *The Reckoning*
- [ ] **QE 85** ($27.99) Sarah Krasnostein *Not Waving, Drowning*
- [ ] **QE 86** ($27.99) Hugh White *Sleepwalk to War*
- [ ] **QE 87** ($27.99) Waleed Aly & Scott Stephens *Uncivil Wars*
- [ ] **QE 88** ($27.99) Katharine Murphy *Lone Wolf*
- [ ] **QE 89** ($27.99) Saul Griffith *The Wires That Bind*
- [ ] **QE 90** ($27.99) Megan Davis *Voice of Reason*
- [ ] **QE 91** ($27.99) Micheline Lee *Lifeboat*
- [ ] **QE 92** ($27.99) Alan Kohler *The Great Divide*
- [ ] **QE 93** ($27.99) Lech Blaine *Bad Cop*

Prices include GST, postage and handling within Australia.
Please include this form with delivery and payment
details overleaf.
Back issues also available as ebooks from ebook retailers.

SUBSCRIBE TO RECEIVE NEARLY 20% OFF THE COVER PRICE

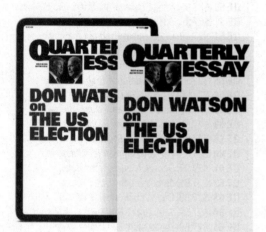

☐ **ONE-YEAR PRINT AND DIGITAL SUBSCRIPTION: $89.99**

- Print edition
- Home delivery
- Automatically renewing
- Full digital access to all past issues
- App for Android and iPhone users
- ebook files

DELIVERY AND PAYMENT DETAILS

DELIVERY DETAILS:

NAME:

ADDRESS:

EMAIL: PHONE:

PAYMENT DETAILS: Enclose a cheque/money order made out to Schwartz Books Pty Ltd.
Or debit my credit card (MasterCard, Visa and Amex accepted).
Freepost: Quarterly Essay, Reply Paid 90094, Collingwood VIC 3066
All prices include GST, postage and handling.

CARD NO. ☐☐☐☐ ☐☐☐☐ ☐☐☐☐ ☐☐☐☐

EXPIRY DATE: /   CCV: AMOUNT: $

PURCHASER'S NAME: SIGNATURE:

Subscribe online at **quarterlyessay.com/subscribe** • Freecall: 1800 077 514 • Phone: 03 9486 0288
Email: subscribe@quarterlyessay.com (please do not send electronic scans of this form)